Wastewater Bacteria

WASTEWATER MICROBIOLOGY SERIES

Editor
Michael H. Gerardi

Wastewater Bacteria

Michael H. Gerardi
Water Pollution Biology
Williamsport, Pennsylvania

WILEY-INTERSCIENCE
A JOHN WILEY & SONS, INC., PUBLICATION

Published by John Wiley & Sons, Inc., Hoboken, New Jersey
Published simultaneously in Canada

For general information on our other products and services or for technical support, please contact
our Customer Care Department within the United States at (800) 762-2974, outside the United States
at (317) 572-3993 or fax (317) 572-4002.

Wiley also publishes its books in a variety of electronic formats. Some content that appears in print
may not be available in electronic formats. For more information about Wiley products, visit our web
site at www.wiley.com.

Library of Congress Cataloging-in-Publication Data:

Gerardi, Michael H.
  Wastewater bacteria / Michael H. Gerardi.
     p. cm.
  Includes bibliographical references and index.
  ISBN-13: 978-0-471-20691-0 (cloth : alk. paper)
  ISBN-10: 0-471-20691-1 (cloth : alk. paper)
1. Sewage sludge digestion.   2. Anaerobic bacteria.   I. Title.
  TD769.G48 2006
  628.3′54—dc22
                                        2005031921

10 9 8 7 6 5 4 3 2 1

To
Rani Harrison

The author extends his sincere appreciation to
joVanna Gerardi for computer support
and
Cristopher Noviello for artwork used in this text.

Contents

Preface

The basic objectives of wastewater treatment are twofold: (1) Degrade organic wastes to a level where they do not exert a significant, dissolved oxygen demand upon receiving waters and (2) remove nutrients (nitrogen and phosphorus) to levels where photosynthetic organisms in receiving waters are limited in their growth. In order to achieve these objectives, it is essential for plant operators to understand the biological processes and organisms involved in wastewater treatment to ensure that the presence of an adequate, active, and appropriate population of bacteria is present in each process. The bacteria are the organisms of primary concern in all biological processes. However, bacteria in wastewater are not a monoculture but, instead, a diversity of organisms that perform different roles and have different operational conditions that are best for their optimal activity and growth (i.e., wastewater treatment).

The large diversity of bacteria and the roles that they perform in wastewater treatment are represented best in two biological treatment units, namely, the activated sludge process and the anaerobic digester. The bacteria and these two biological treatment units are reviewed in this book. The activated sludge process is the most commonly used aerobic biological treatment unit at municipal wastewater treatment plants. The organisms here consist of procaryotes (bacteria) and eucaryotes (protozoa and metazoa). The biological processes occur in aerobic and anoxic environments and are based on respiration. The anaerobic digester is the most commonly used anaerobic biological treatment unit at municipal wastewater treatment plants. The organisms consist exclusively of procaryotes. The biological processes occur in an anaerobic environment and are based on fermentation. There are significant differences in the microbial communities between the activated sludge process and the anaerobic digester.

This book reviews the significant bacterial groups, the roles they perform in wastewater treatment, and the operational conditions that affect their activity. The roles that are performed by each bacterial group may be beneficial or detrimental to the biological treatment unit and depend upon the operational conditions of the unit. Effective control and proper operation of each biological treatment unit is based upon an understanding of the basic principles of bacterial activity and growth that are presented in this book.

Several of the significant groups of bacteria that are reviewed in this book are denitrifying bacteria, fermentative (acetate-forming and acid-forming) bacteria, filamentous bacteria, floc-forming bacteria, hydrolytic bacteria, methane-forming bacteria, nitrifying bacteria, poly-P bacteria, sulfur-oxidizing bacteria, and sulfur-reducing bacteria. Several of these bacterial groups are presented in comprehensive reviews in other books in the Wastewater Microbiology Series.

Wastewater Bacteria is the fifth book in the Wastewater Microbiology Series by John Wiley & Sons. This series is designed for wastewater personnel, and the series presents a microbiological review of the significant groups of organisms and their roles in wastewater treatment facilities.

MICHAEL H. GERARDI
State College, Pennsylvania

Part I

Bacteria and Their Environment

1

Wastewater Microorganisms

Although most organisms in biological wastewater treatment plants are microscopic in size, there are some organisms such as bristleworms and insect larvae that are macroscopic in size. Macroscopic organisms can be observed with the naked eye—that is, without the use of a light microscope. Microscopic organisms can only be observed with the use of a light microscope. Of the microscopic organisms the bacteria (singular: bacterium) are the most important in wastewater treatment plants and can be seen with the light microscope only under highest magnification. Several groups of microorganisms such as protozoa and some metazoa possess large and more complex cells that can be observed easily with the light microscope without the use of highest magnification. Compared to other organisms, microorganisms have relatively simple structures.

All living cells can be classified as procaryotic or eucaryotic (Table 1.1). Procaryotic cells lack a nucleus and other membrane-bound structures, while eucaryotic cells possess these structures (Figure 1.1). The nucleus is the primary membrane-bound structure in eucaryotic cells. It regulates cellular activity and contains the genetic information. Examples of membrane-bound structures or organelles found in eucaryotic cells include the golgi apparatus (which regulates cellular metabolism) and lysomes (which contain hydrolytic enzymes).

Based upon cellular structure and function, microorganisms are commonly classified as eucaryotes and procaryotes. The procaryotes consist of (1) eubacteria or "true" bacteria and (2) archaebacteria or "ancient" bacteria (Table 1.2). The eubacteria and archaebacteria are the most important microorganisms in biological, wastewater treatment plants. Together, these two procaryotes commonly are referred to as bacteria.

Wastewater Bacteria, by Michael H. Gerardi
Copyright © 2006 John Wiley & Sons, Inc.

TABLE 1.1   Major Differences between Procaryotic Organisms and Eucaryotic Organisms

Feature	Procaryotic Organism	Eucaryotic Organism
Genetic material	Not contained in a membrane	Contained in a membrane
Organelles	None	Many
Structure	Simple	Complex

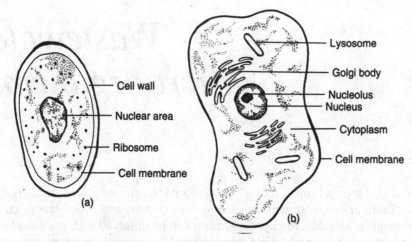

FIGURE 1.1   *Procaryotic and eucaryotic cells. The procaryotic cell (a) contains no membrane-bound organelles such as the nucleus, golgi body, and lysosome that are found in the eucaryotic cell (b).*

TABLE 1.2   Classification of Microorganisms in Wastewater Treatment Plants

Group	Cell Structure	Organization	Representatives
Eucaryotes	Eucaryotic	Multicellular	Bristleworms, flatworms, free-living nematodes, waterbears
Eubacteria	Procaryotic	Unicellular	Bacteria
Archaebacteria	Procaryotic	Unicellular with unique cellular chemistry	Halophiles, methanogens, thermacidophiles

There are four important eucaryotic organisms in the activated sludge process. These organisms are fungi, protozoa, rotifers, and nematodes. These free-living (non-disease-causing) eucaryotes enter wastewater treatment plants through inflow and infiltration (I/I) as soil and water organisms.

FUNGI

Fungi usually are saprophytic organisms and are classified by their mode of reproduction. As saprophytes they obtain their nourishment from the degradation of dead organic matter. Most fungi are free-living and include yeast, molds, and mushrooms.

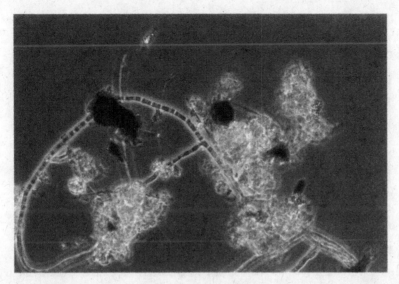

FIGURE 1.2 *Filamentous fungi. Filamentous fungi occasionally bloom in activated sludge processes due to low pH or nutrient deficiency. Filamentous fungi are relatively large in size and display true branching.*

Most fungi are strict aerobes and can tolerate a low pH and a low nitrogen environment. Although fungi grow over a wide range of pH values (2–9), the optimum pH for most species of fungi is 5.6, and their nitrogen nutrient requirement for growth is approximately one-half as much as that for bacteria.

In the activated sludge process filamentous fungi (Figure 1.2) may proliferate and contribute to settleability problems in secondary clarifiers. The proliferation of filamentous fungi is associated with low pH (<6.5) and low nutrients. Although filamentous fungi contribute to settleability problems in the activated sludge process, the presence of a large and diverse population of fungi is desired for the treatment of some industrial wastewaters and composting of organic wastes. Fungi have the ability to degrade cellulose, tolerate low nutrient levels, and grow in the presence of low moisture and low pH conditions.

An example of a unicellular fungus is the yeast (*Saccharomyces*). They reproduce by budding. Budding results in the production of numerous daughter cells (offspring) from one parent cell. Yeast can degrade organic compounds to carbon dioxide and water with the use of free molecular oxygen (O_2), or as facultative anaerobes they can degrade organic compounds such as sugars to ethanol (CH_3CH_2OH) in the absence of free molecular oxygen.

PROTOZOA

Protozoa are unicellular organisms. Most protozoa are free-living and solitary, but some do form colonies. Most protozoa are strict aerobes, but some including amoebae and flagellates can survive anaerobic conditions.

In the activated sludge process, protozoa are placed commonly in five groups according to their means of locomotion. These groups are amoebae (Figure 1.3),

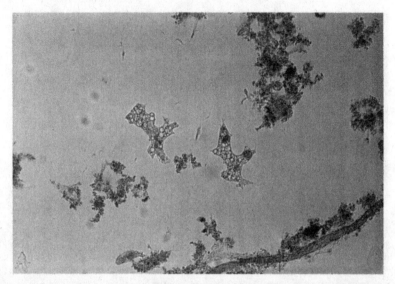

FIGURE 1.3 Amoeba. The amoeba is a single-celled organism that moves by a pseudopodia ("false-foot") mode of locomotion—that is, the streaming of cytoplasm against the cell membrane.

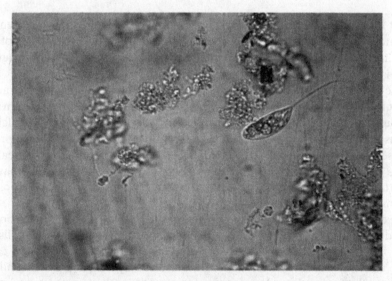

FIGURE 1.4 Flagellate. The flagellate is a single-celled organism that moves by the beating action of one (flagellum) or more (flagella) whip-like structures.

flagellates (Figure 1.4), free-swimming ciliates (Figure 1.5), crawling ciliates (Figure 1.6), and stalked ciliates (Figure 1.7).

Ciliated protozoa are the most important groups of protozoa in the activated sludge process. They possess short hair-like structures or cilia that beat in unison to produce a water current for locomotion and food gathering—that is, to bring bacteria into their mouth opening (Figure 1.8). Ciliated protozoa provide the following benefits to the activated sludge process:

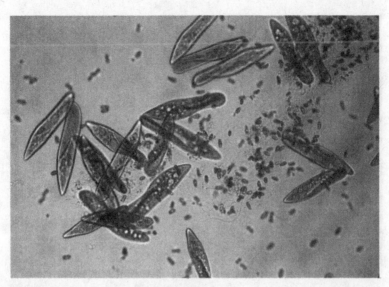

FIGURE 1.5 *Free-swimming ciliate. The free-swimming ciliate is a single-celled organism that moves by the beating action of hair-like structures or cilia that are found in rows that cover the entire surface of the organism.*

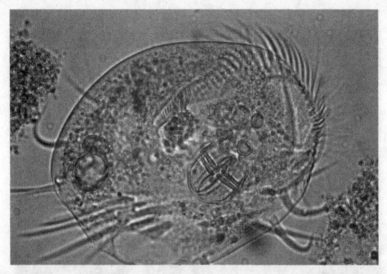

FIGURE 1.6 *Crawling ciliate. The crawling ciliate is a single-celled organism that moves by the beating action of hair-like structures or cilia that are found in rows that cover only the ventral or "belly" surface of the organism.*

- Add weight to floc particles and improve their settleability.
- Consume dispersed cells and cleanse the waste stream.
- Produce and release secretions that coat and remove fine solids (colloids, dispersed cells, and particulate material) from the bulk solution to the surface of floc particles.
- Recycle nutrients (nitrogen and phosphorus) through their excretions.

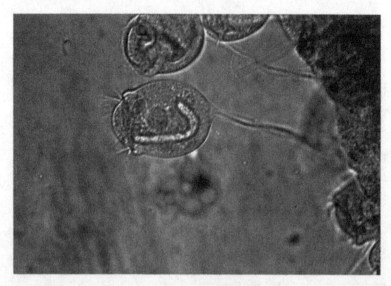

FIGURE 1.7 *Stalked ciliate. The stalked ciliate is a single-celled organisms that moves by the beating action of hair-like structures or cilia that are found in rows that surround only the mouth opening of the organism. Some stalked ciliates may grow in a colony, and some by "spring" by means of a contractile filament or myoneme in the posterior portion or "stalk" of the organism.*

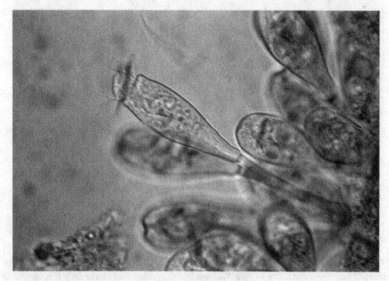

FIGURE 1.8 *Cilia surrounding the mouth opening of a stalk ciliated protozoa.*

ROTIFERS AND NEMATODES

Rotifers (Figure 1.9) and nematodes (Figure 1.10) are multicellular microscopic animals (metazoa) that also provide numerous benefits to the activated sludge process. In addition to these benefits provided by the ciliated protozoa, the metazoa burrow into floc particles. The burrowing action promotes acceptable bacterial activ-

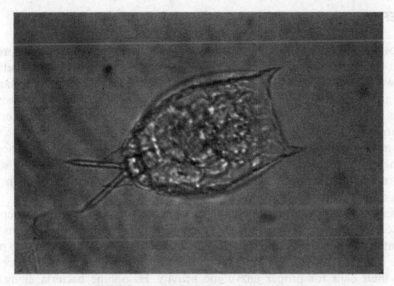

FIGURE 1.9 *Rotifer in free-swimming mode.*

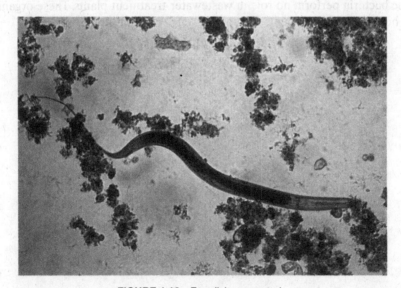

FIGURE 1.10 *Free-living nematode.*

ity for the degradation of substrates in the core of the floc particle by permitting the penetration of dissolved oxygen, nitrate (NO_3^-), substrates, and nutrients. Substrates are the carbon and energy sources used by bacteria for cellular growth and activity. With exceptions, substrates consist of carbonaceous, biochemical oxygen demand (cBOD) compounds and nitrogenous, biochemical oxygen demand (nBOD) compounds.

BACTERIA

The most important organisms in biological, wastewater treatment plants are the bacteria—eubacteria and archaebacteria. Recognition of the distinction between these two groups of organisms or domains (Bacteria and Archaea) is relatively recent, and it is common for species of both groups to be referred to as bacteria. Bacteria enter wastewater treatment plants through fecal waste and I/I as soil and water organisms.

The archaebacteria consist of the halophiles, thermacidophiles, and methanogens. Only the methanogens or methane-forming bacteria are of importance in wastewater treatment plants. Methane-forming bacteria stabilize wastes through their conversion to methane (CH_4).

Halophiles (salt-loving) or halophilic bacteria are found in saltwater where the salt concentration (3.5%) is optimum for their growth. These marine organisms need an elevated sodium ion (Na^+) concentration in their environment in order to maintain the integrity of their cell wall and an elevated potassium ion (K^+) concentration in their cells for proper enzymatic activity. Halophilic bacteria along with cyanobacteria and photosynthetic bacteria produce gas vacuoles. These vacuoles are used to regulate cell buoyancy; that is, they are a cellular floatation device.

Thermacidophiles (high-temperature-loving and low-pH-loving) or thermacidophilic bacteria perform no role in wastewater treatment plants. These organisms live in hot acidic environments such as volcanic vents on the ocean floor.

2

Microbial Ecology

Microbial ecology as applied to the activated sludge process and the anaerobic digester is the review of the significant groups of wastewater organisms and the operational conditions in each biological treatment unit. This review includes the effects of abiotic and biotic factors upon the organism including their activity and growth—that is, wastewater treatment efficiency. Biological treatment units are simply biological amplifiers—that is, the removal or degradation of waste results in an increase in the number of organisms (sludge). Therefore, acceptable activity and growth of the organisms or biomass is acceptable wastewater treatment.

Collectively, all organisms and operational conditions are interrelated by the transfer of carbon and energy through a food chain (Figure 2.1) or more appropriately a food web (Figure 2.2). Within the food web there are numerous habitats, niches, and relationships (symbiotic and predator–prey) that determine the success or failure of the biological treatment unit to treat wastewater.

Abiotic factors are the nonliving components or operational conditions in a biological treatment unit that affect the activity and growth of the biomass. For example, a decrease in pH of the activated sludge process favors the proliferation of filamentous fungi and disfavors the growth of bacteria, and a decrease in pH in the anaerobic digester favors the growth of fermentative bacteria and disfavors the growth of methane-forming bacteria. Biotic factors are the living components or organisms in a biological treatment unit. Each organism has an effect upon other organisms (predator–prey and symbiotic relationships) and abiotic factors in the biological treatment unit. For example, free-swimming ciliated protozoa increase in number in the presence of large numbers of dispersed bacterial cells. However, during floc formation the number of dispersed bacterial cells decreases and, consequently, the number of free-swimming ciliated protozoa decrease in number. In the

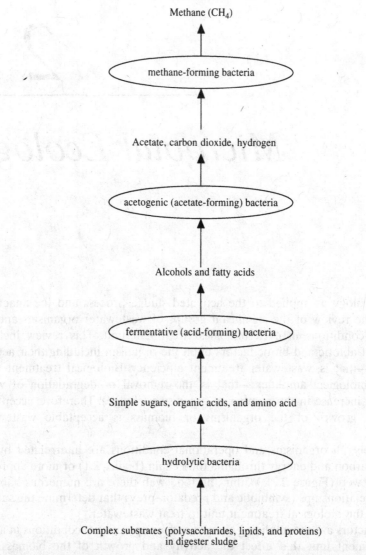

Methane (CH$_4$)

↑

methane-forming bacteria

↑

Acetate, carbon dioxide, hydrogen

↑

acetogenic (acetate-forming) bacteria

↑

Alcohols and fatty acids

↑

fermentative (acid-forming) bacteria

↑

Simple sugars, organic acids, and amino acid

↑

hydrolytic bacteria

↑

Complex substrates (polysaccharides, lipids, and proteins)
in digester sludge

FIGURE 2.1 *Transfer of carbon and energy through an anaerobic digester food chain. Carbon and energy enter the anaerobic digester in the form of large, complex organic molecules such as polysaccharides, lipids, and proteins. These compounds are degraded to smaller and simpler compounds through step-by-step biochemical reactions by a diversity of bacterial groups to methane. Through each biochemical reaction, bacteria are produced.*

activated sludge process, nitrifying bacteria decrease alkalinity and pH, while denitrifying bacteria increase alkalinity and pH.

In the anaerobic digester, four different groups of bacteria have a symbiotic relationship (Figure 2.3). Fermentative bacteria increase the quantities of carbon dioxide and hydrogen (H$_2$), while hydrogenotrophic methane-forming bacteria decrease the quantities of carbon dioxide and hydrogen. Hydrogenotrophic methane-forming bacteria combine carbon dioxide and hydrogen to form methane. By using

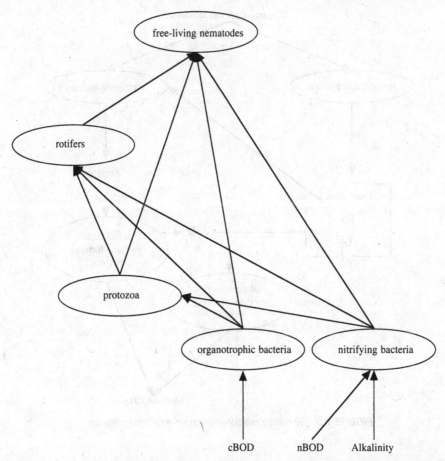

FIGURE 2.2 *Transfer of carbon and energy through an activated sludge food web. Carbon and energy enter the activated sludge process in the form of cBOD and nBOD and alkalinity. These carbon and energy substrates are used by a variety of organisms in the activated sludge process, and many of the organisms that grow from these substrates in turn are used as substrates by other organisms. The transfer of carbon and energy in the activated sludge process is between many groups of organisms in a "web-like" pattern.*

hydrogen to produce methane, the hydrogen pressure in the anaerobic digester decreases. This decrease in hydrogen pressure enables acetogenic bacteria to produce acetate (CH_3COOH). Acetoclasitc methane-forming bacteria use acetate to produce methane and carbon dioxide. The hydrogenotrophic methane-forming bacteria also combine the carbon dioxide produced by the acetogenic bacteria with hydrogen to form methane.

However, when the hydrogenotrophic methane-forming bacteria are inhibited, the hydrogen pressure increases in the anaerobic digester. The increase in hydrogen pressure inhibits acetogenic bacteria. This results in a decrease in acetate production and consequently a decrease in methane production by acetoclastic methane-forming bacteria.

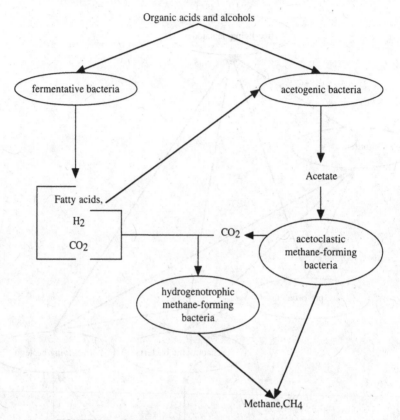

FIGURE 2.3   Symbiotic relationships in an anaerobic digester.

ACTIVATED SLUDGE PROCESS, SIGNIFICANT ABIOTIC AND BIOTIC FACTORS

Significant abiotic factors in the activated sludge process include alkalinity, ionized ammonia (NH_4^+), dissolved oxygen, hydraulic retention time (HRT), nutrients, pH, quantity and types of substrates, return activated sludge (RAS) rate, temperature, toxic wastes, and turbulence. Significant biotic factors include denitrifying bacteria, filamentous organisms, floc-forming bacteria, mean cell residence time (MCRT) or sludge age, mixed liquor volatile suspended solids (MLVSS) concentration, nitrifying bacteria, and relative abundance and dominant groups of protozoa.

ANAEROBIC DIGESTER, SIGNIFICANT ABIOTIC AND BIOTIC FACTORS

Significant abiotic factors in the anaerobic digester include ionized ammonia (NH_4^+), alkalinity, carbon dioxide, hydrogen, nitrate (NO_3^-), nutrients, pH, quantity and types of substrates, sulfate (SO_4^{2-}), temperature, toxic wastes, and volatile acids. Significant biotic factors include acetogenic bacteria, fermentative (acid-forming) bacteria, hydrolytic bacteria, methane-forming bacteria, solids retention time (SRT), sulfur-reducing bacteria, and volatile suspended solids (VSS).

Within each biological treatment unit, different groups of organisms transfer carbon and energy from one trophic (food) level to the next trophic level (Figures 2.1 and 2.2). In the activated sludge process, carbon and energy enter the process as nonliving substrates or BOD. In the soluble form, BOD is absorbed by a variety of organisms, mostly bacteria. Some of the absorbed BOD is transformed into new bacterial cells (sludge) or living BOD. Each organism in the food chain or food web represents BOD, because living organisms are consumed (predator–prey relationships); for example, bacteria are consumed by protozoa and metazoa, and dead organisms are decomposed by living organisms.

As carbon and energy move up the food chain or food web, the quantity (weight) or biomass of each group of organism in the higher trophic level decreases (Figure 2.4). With each move to a higher trophic level, more carbon and energy are lost in waste products and heat, thus leaving less carbon and energy for the synthesis of cellular material (biomass). However, the transfer of carbon and energy from one group of organisms to another is not as simple as a food chain, because several groups of organisms often feed off the same substrates or lower trophic level. The transfer or movement of carbon and energy here is referred to as a food web. The food web better illustrates the activated sludge process than the food chain, because organisms here work mostly side-by-side (Figure 2.2). For example, nitrifying bacteria oxidize nBOD, while organotrophic bacteria oxidize cBOD, and biological phosphorus removal occurs, while organotrophic bacteria oxidize cBOD.

The food chain better illustrates the anaerobic digester than the food web, because bacteria in the digester work in step-by-step fashion to produce methane

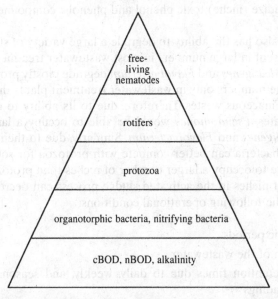

FIGURE 2.4 *Activated sludge food pyramid. As carbon and energy move up the food pyramid from one tropic level to the next, a smaller quantity of organisms (biomass) is produced in the higher trophic level, due to the loss of some carbon and energy as carbon dioxide and waste products. The loss occurs as a result of the biochemical reactions involved in the degradation of the substrate (biomass) from the lower trophic level.*

(Figure 2.1). Hydrolytic bacteria solubilize complex organic wastes to soluble wastes. The soluble wastes then are degraded to simple wastes by fermentative bacteria. These simple wastes finally are converted to methane by methane-forming bacteria.

HABITATS AND NICHES

Each group of organisms in a biological treatment unit has its own habitat and niche. Changes in operational conditions can decrease or increase the number of niches. The habitat is where the organism lives, and the niche is the role the organism performs in the biological treatment unit. Examples of habitats and niches are numerous in the activated sludge process and the anaerobic digester.

Examples of habitats and niches in the activated sludge process include the following:

- Strict aerobic nitrifying bacteria live (habitat) mostly on the perimeter of the floc particle where dissolved oxygen concentration is high, and they oxidize (niche) ionized ammonia and nitrite (NO_2^-).
- Floc-forming bacteria live (habitat) throughout the floc particle and agglutinate (niche) bacteria and remove (niche) fine solids and heavy metals from the waste stream. Floc-forming bacteria also oxidize (niche) soluble cBOD.
- Pseudomonads (members of the genus *Pseudomonas*) are found (habitat) throughout the floc particle and oxidize (niche) soluble cBOD in the presence or absence of free molecular oxygen. Pseudomonads also have the unique ability to oxidize (niche) toxic phenol and phenolic compounds.

Pseudomonas also has the ability to degrade a large variety of substrates; therefore, they are present in large numbers in most wastewater treatment plants. Other bacteria such as *Alcaligenes* and *Flavobacterium* degrade mostly proteins and would be present in large numbers only in wastewater treatment plants that receive large amounts of proteinaceous wastes. Therefore, due to its ability to degrade a large variety of substrates, *Pseudomonas* would be able to occupy a larger number of niches than *Alcaligenes* and *Flavobacterium*. Similarly, due to their larger surface-to-volume ratio, bacteria can better compete with protozoa for soluble substrates and would be able to occupy a larger number of niches than protozoa.

The number of niches in the activated sludge process can decrease or increase with changes in the following operational conditions:

- Use of anoxic periods
- Composition of the wastewater
- Hydraulic retention times due to daily, weekly, and seasonal variations in hydraulic loadings
- Return activated sludge rates

Examples of habitats and niches in the anaerobic digester also are numerous and include the following:

- Anaerobic hydrolytic bacteria live (habitat) in a redox environment $<-100\,mV$ and solubilize (niche) complex organic compounds.
- Sulfate-reducing bacteria live (habitat) in a redox environment $<-100\,mV$ and reduce (niche) sulfate (SO_4^{2-}) to sulfides (H_2S and HS^-) and make (niche) sulfur available as a nutrient to anaerobic bacteria.
- Fermentative (acid-forming) bacteria live (habitat) in a redox environment $<-200\,mV$ and produce (niche) substrates that can be used by methane-forming bacteria.
- Methane-forming bacteria live (habitat) in a redox environment $<-300\,mV$ and remove (niche) carbon dioxide, hydrogen, and acetate from the environment.

The number of niches in the anaerobic digester also can decrease or increase with changes in the following operational conditions:

- Composition of the substrates transferred to the anaerobic digester
- Temperature
- pH and the form or toxic wastes rendering them more or less toxic
- Hydrogen pressure

COMPETITION

Competition for a niche is continuous and is a critical factor in determining the success of a biological treatment unit. Competition is better known as the "competitive exclusion principle." This principle states that only one species can occupy any niche at any time. The species that occupies the niche is the species that can utilize the resources of the environment most efficiently.

In the activated sludge process, filamentous organisms such as type 1701 and type 021N proliferate under a low nutrient condition for nitrogen or phosphorus, because they can better compete for nutrients than most floc bacteria. The competitive advantages of the filamentous organisms to obtain nutrients when they are limited in quantity are twofold. First, the filamentous organisms have more surface area exposed to the bulk solution than floc bacteria do. Therefore, they are better able to absorb larger quantities of nutrients. Second, the filamentous organisms simply require a smaller quantity of nutrients, and their growth is not inhibited in the presence of relatively low concentrations of nutrients. Similarly, filamentous organisms such as *Sphaerotilus natans* and *Haliscomenobacter hydrossis* proliferate under a low dissolved oxygen concentration, because they can better compete for the limited quantity of dissolved oxygen than most floc bacteria.

With increasing sulfate (SO_4^{2-}) concentration in an anaerobic digester, sulfate-reducing bacteria compete with methane-forming bacteria for some identical substrates. Because sulfate-reducing bacteria are more active than methane-forming bacteria, they can better compete for the substrates that are needed by the methane-forming bacteria. This competition results in a decrease in the activity and growth of methane-forming bacteria and a decrease in methane production.

SYMBIOTIC RELATIONSHIPS

Symbiotic relationships are beneficial relationships between organisms. Symbiotic relationships are important in the treatment of wastewater in the activated sludge process and sludge in the anaerobic digester. An example of a symbiotic relationship in the activated sludge process is the oxidation of ionized ammonia by *Nitrosomonas* to nitrite. The oxidation of ionized ammonia benefits *Nitrobacter* by providing *Nitrobacter* with the nitrite energy substrate.

An example of a symbiotic relationship in the anaerobic digester involves three different groups of bacteria. Hydrogenotrophic methane-forming bacteria remove hydrogen (H_2) from the digester to produce methane (CH_4). By removing hydrogen, the hydrogen pressure in the digester remains low. The low hydrogen pressure permits acetogenic bacteria to produce acetate (CH_3COOH). The production of acetate provides substrate for acetoclastic methane-forming bacteria.

PREDATOR–PREY RELATIONSHIPS

Numerous predator–prey relationships exist in the activated sludge process. Although bacteria can better compete for soluble substrates than protozoa, the bacteria serve as particulate substrate (prey) for protozoa and metazoa (predators). The protozoa, too, serve as particulate substrate (prey) for the metazoa (predator). Metazoa may not be present in large numbers in the activated sludge process, if the MCRT is <28 days due to the long generation time of most metazoa.

Changes in abiotic and biotic factors in a biological treatment unit change the number of niches and the dominant organisms in the microbial community as well as the relationships between groups of organisms. These changes may have a positive or negative impact upon the activity and growth of the bacteria. Therefore, it is important for operators of wastewater treatment plants to be aware of the complexities of microbial ecology and the need to identify, monitor, and regulate those operational conditions that are most critical for the success of each biological treatment unit.

3

Bacteria

Bacteria are the simplest life form and are the most numerous organisms with respect to number of species and total biomass. They are small, unicellular procaryotic organisms (Figure 3.1). Bacteria are classified by structure (morphology), response to chemical stains, nutrition, and metabolism.

Except for filamentous forms, cyanobacteria, and spirochetes (Figure 3.2), the range in sizes of most bacteria is from 0.3 to 3 µm. Filamentous bacteria such as *Sphaerotilus natans* usually are >100 µm in length. The unicellular cyanobacteria have photosynthetic pigments in cell membranes and range in size from 5 to 50 µm. Free-living spirochetes commonly are found in wastewater treatment plants and may be up to 50 µm in length. Although the size range for bacteria is described as a diameter or largest side of the bacterial cell, there are many bacteria that are not spherical. The size of these organisms may be described according to their length and width. For example, *Escherichia coli*, a common bacterium found in human feces and wastewater treatment plants, is approximately 2 µm in length and 0.5 µm in width.

Young bacterial cells of the same species are smaller than old bacterial cells of the same species and have a higher growth rate. Young bacterial cells increase in size only to reproduce. The higher growth rate of young bacterial cells as compared to old bacterial cells is due to the larger surface-to-volume ratio of young bacterial cells. The larger surface-to-volume ratio provides more surface area for the absorption of substrates and nutrients and, consequently, a higher metabolic rate of activity including growth and reproduction.

Most bacteria can be grouped into three basic shapes: bacillus (rod), coccus (sphere), and spirillum (spiral) (Figure 3.3). In some species of bacteria the offspring or daughter cells do not separate after division. The lack of separation results in the formation of several arrangements of bacterial growth including colonies, tetrads, and chains or filaments (Figure 3.3).

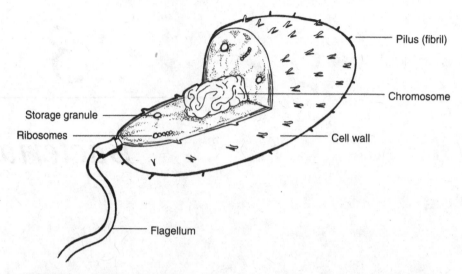

Pilus (fibril)

Chromosome

Storage granule

Ribosomes

Cell wall

Flagellum

FIGURE 3.1 *Typical procaryotic cell. Diagram of a typical procaryotic cell or bacterium.*

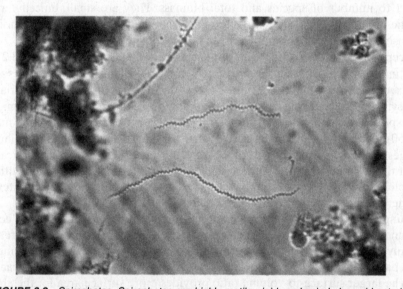

FIGURE 3.2 *Spircohetes. Spirochetes are highly motile, rigid, and spiral-shaped bacteria.*

Coccus-shaped bacteria can divide in one or two planes. Division in one plane produces cells in pairs (diplococcus) or chains (streptococcus). Division in two planes produces cells in grape-like clusters or tetrads. Bacillus-shaped bacteria divide in only one plane and produce cells in chains that are placed end-to-end or side-by-side. Spirillum-shaped bacterial cells usually are not found in multicellular arrangements. Bacteria typically reproduce by binary fission in which a mother cell divides into two equal daughter cells.

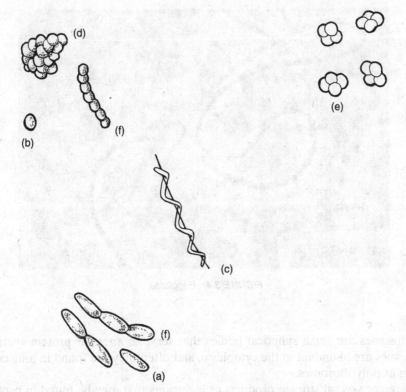

FIGURE 3.3 *Bacterial shapes and growth patterns. Basic bacterial shapes consist of bacillus (a), coccus (b), and spriillum (c). Due to the lack of separation of bacterial cells after reproduction, several basic patterns of growth result. These patterns include colonies (d), tetrads (e), and chains or filaments (f).*

CELL STRUCTURE

Structurally, the bacterial cell can be divided into the following (Figure 3.1):

- The cytoplasm and its contents
- The cell membrane, cell wall, and outer membrane
- A variety of external structures

Cytoplasm and Its Contents

The cytoplasm is inside the cell membrane and is mostly water by composition. However, it has a semifluid nature due to a suspension of carbohydrates, enzymes, inorganic ions, lipids, and proteins. Within this suspension can be found the nuclear region, ribosomes, storage products, and endospores.

Instead of a nucleus, bacterial cells have a nuclear region or nucleoid. The nuclear region consists mostly of genetic material in one large and circular chromosome. In addition to the chromosome, some bacteria have small molecules of genetic material, called plasmids, that contain additional genetic information.

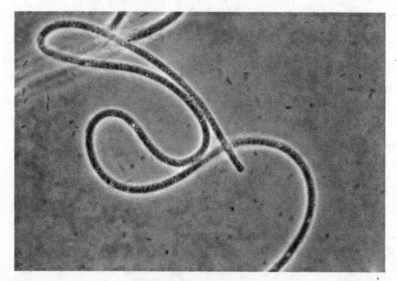

FIGURE 3.4 Beggiatoa.

Ribosomes are small spherical bodies that serve as sites for protein synthesis. Ribosomes are abundant in the cytoplasm, and often they are found in long chains known as polyribosomes.

There are several storage products or inclusions that may be found in bacterial cells. These products include stored food, volutin granules, and sulfur granules. Food may be stored in the form of glycogen (glucose polymer), starch, and poly-β-hydroxybutyrate (PHB). Phosphorus-accumulating organisms or poly-P bacteria including *Acinobacter* store phosphorus in long chains of polyphosphate known as volutin granules. Volutin granules also are known as metachromatic granules, because they display metachromasia. Metachromasia is not a stain of uniform color produced by a simple stain. It is the production of different intensities of color. Sulfur-oxidizing bacteria including the filamentous organisms *Beggiatoa* (Figure 3.4) and *Thiothrix* oxidize sulfides (HS^-) as an energy source. The product from the oxidation of sulfide is insoluble elemental sulfur ($S°$) that is stored as a granule in the cytoplasm.

Endospores are produced by only a small number of bacteria including species of *Bacillus* and *Clostridium*. Endospores are dormant reproduction bodies of active vegetative bacteria. They contain very little water and are highly resistant to harsh environmental conditions including acidity, desiccation, heat, and some disinfectants. When conditions become favorable, the endospores germinate.

Cell Membrane, Cell Wall, and Outer Membrane

The cell membrane or plasma membrane is a flexible semipermeable membrane that surrounds the cytoplasm. The cell membrane contains two different layers (bilayer) that regulate the movement of substances in and out of the cell. The outer layer is hydrophilic (water-loving), while the inner layer is hydrophobic (water-fearing). Together these layers form a protective and regulating barrier between the cytoplasm and the environment.

TABLE 3.1   Significant Differences Between Gram-Negative and Gram-Positive Bacteria

Characteristic	Gram-Negative Bacteria	Gram-Positive Bacteria
Lipids	Much lipopolysaccharide	Very little
Peptidoglycan	Thin layer	Thick layer
Outer membrane	Present	Absent

FIGURE 3.5 *Axial filament in spirochete.*

The cell wall is an extremely porous and rigid structure that lies outside the cell membrane. The cell wall performs three functions. First, it provides protection. Second, it maintains the characteristic shape of the cell. Third, it prevents the cell from bursting when fluids flow into the cytoplasm. Based upon chemical composition, there are two types of cell walls, Gram negative and Gram positive (Table 3.1).

Gram-positive bacteria retain the crystal violet stain after the Gram-staining technique due to their thick cell wall. Gram-negative bacteria cannot retain the crystal violet stain due to their thin cell wall and presence of relatively large quantities of lipoproteins and lipopolysaccharides. Gram negative bacteria are stained red with Safranin at the end of the Gram staining technique.

An outer membrane is found only in Gram-negative bacteria. It is a bilayer membrane and surrounds the cell wall. The outer membrane controls the transportation of some proteins from the environment into the cell.

External Structures

There are several external structures. These structures are the axial filament, capsule, flagella, glycocalyx, and pili. All external structures perform specific functions, but these structures are not found on bacteria.

Spirochetes possess axial filaments or endoflagella (Figure 3.5). The axial filament runs the length of the spirochete and extends beyond the cell wall. The filament assists provides motility and allows the rigid spirochete to rotate like a corkscrew.

The capsule is a protective structure that is located outside the cell wall. The capsule consists of polysaccharides that are arranged in a loose gel. Very few bacteria produce a capsule.

Nearly half of all bacteria are motile. Except for the axial filament, most bacteria move by means of flagella (Figure 3.6). Each flagellum is a long, helical structure that is made of protein subunits call flagellin. The beating action or whip-like motion of the flagella propels the bacterium.

Glycocalyx is the term used to describe all polysaccharides outside the cell wall. Glycocalyx may be packaged as a thick capsule or loosely collected as a thin slime layer. Glycocalyx enables bacteria to flocculate and form (1) floc particles in the

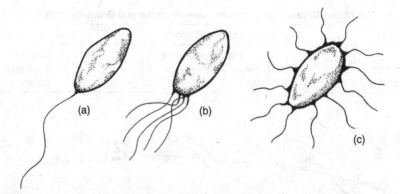

FIGURE 3.6 *Flagellar arrangements in bacteria. The three basic flagellar arrangements in bacteria are monotrichous (a), lophotrichous (b), and peritrichous (c).*

activated sludge process and (2) biofilm in fixed film processes such as the trickling filter.

Pili or fibrils are short hollow projections of the cell membrane that extend through the cell wall into the surrounding environment. Each pilus is composed of protein subunits called pilin. The pili are used by some bacteria for attachment to other bacteria for reproduction and by many bacteria for attachment to other bacteria or inert surfaces (floc formation and biofilm development).

CELLULAR COMPOSITION

Bacteria are perhaps the most versatile and diversified organisms with regard to their nutritional requirements. Some represent the entire spectrum of nutritional types, while others require complex organic compounds or a few inorganic compounds.

Although there is considerable variation in the specific requirements for growth, the chemical composition of bacteria is similar (Table 3.2). Bacteria are approximately 80% water and 20% dry material. Of the dry material, approximately 90% is organic and 10% is inorganic. A simple organic formula for a bacterial cell that includes nitrogen is $C_5H_7O_2N$. Although the inorganic composition of bacterial cells is relatively small, a shortage of any inorganic element can limit bacterial growth and wastewater treatment plant efficiency.

The major elements (macroelements) in the composition of bacterial cells include C, H, N, O, P, and S. These elements are required in large quantities. The minor elements (microelements) such as Ca, Fe, K, Mg, and Na are required in small quantities, and the trace elements including Co, Mn, Mo, Ni, and Zn are required in relatively small quantities for most bacteria. Some bacteria such as methane-forming bacteria require sulfur and some minor and trace elements such as Co, Fe, and Ni in quantities 2–5 times greater than other bacteria, while halophiles require large quantities of chlorine (Cl), and sodium, and bacteria that synthesize vitamin B_{12} require Co in large amounts. Calcium is required in large amounts by Gram-positive bacteria for the synthesis of cell walls.

TABLE 3.2   Typical Composition of Bacterial Cells (Dry Weight)

Element	Average Composition (%)	Range in Composition (%)[a]
Carbon	50	45–55
Oxygen	20	16–22
Nitrogen	14	12–16
Hydrogen	8	7–10
Phosphorus	3	2–5
Sulfur	1	0.8–1.5
Potassium	1	
Sodium	1	
Calcium	0.5	
Chlorine	0.5	
Magnesium	0.5	
Iron	0.2	
Trace elements	0.1	

[a] Ranges in composition of elements are provided only for the major elements.

MAJOR ELEMENTS

Nearly all bacteria obtain carbon from organic compounds (heterotrophs) or carbon dioxide (autotrophs). Oxygen and hydrogen requirements for cellular synthesis are often satisfied together by the availability of organic compounds.

Nitrogen is available for bacterial use in several compounds. Inorganic compounds such as ionized ammonia (NH_4^+), ammonium salts, nitrate (NO_3^-), and nitrite (NO_2^-) are most often used. Some bacteria including a small group of methane-forming bacteria use atmospheric or molecular nitrogen (N_2).

Sulfur is available to aerobic and facultative anaerobic bacteria in the oxidized form as sulfate (SO_4^{2-}). Sulfur is available to anaerobic bacteria in the reduced form as sulfide (HS^-). Some bacteria are capable of using sulfur-containing amino acids as a source of sulfur. Phosphorus is available to bacteria as phosphate. The form of phosphate ($H_2PO_4^-$, HPO_4^{2-}, or PO_4^{3-}) used by bacteria is pH-dependent.

The growth of bacteria in wastewater treatment plants is affected by many factors including the presence of available nutrients (major elements, minor elements, and trace elements). Consequently, nutrient addition to biological treatment units may be required when soluble, cBOD-rich industrial wastewaters are being treated. These wastewaters often are nutrient deficient. The most commonly occurring deficiencies for nutrients in industrial wastewaters are the major elements nitrogen and phosphorus, while deficiencies for minor and trace elements (calcium, cobalt, iron, and nickel) do occur.

TRANSPORT MECHANISMS

Bacteria have an absorptive means of nutrition; that is, substrates and nutrients must be water-soluble or lipid-soluble in order to enter the bacterial cell. Inside the cell, biochemical reactions using substrates and nutrients occur in water. Waste products from these biochemical reactions are eliminated from the cell in water.

In order for substrates and nutrients to enter the cell and for wastes to leave the cell, they must cross the cell membrane. The cell membrane is semipermeable and contains a layer of lipids that are hydrophobic and influence the mechanism and rate that substances cross the membrane.

There are three basic mechanisms that are used by bacteria to transport substrates, nutrients, and wastes across the cell membrane. These mechanisms are diffusion, facilitated diffusion, and active transport.

Due to the presence of a lipid layer in the cell membrane, lipids and non-ionized substances diffuse better through the membrane than do ionized substances. The rate of diffusion of substrates, nutrients, and wastes is determined by their concentrations inside and outside the cell and their degree of compatibility of movement through the lipid layer.

Diffusion is the random movement of substances from regions of high concentration to regions of low concentration. Many toxic wastes enter bacterial cells through diffusion. Diffusion does not require an expenditure of cellular energy. Diffusion across a semipremeable membrane is referred to as osmosis.

A few substances also diffuse through pores in the cell membrane. This diffuse is affected by the size and charge of the diffusing substance and the charge on the pore surface. Pores in the cell membrane are approximately $0.008\,\mu m$ in size. Therefore, only water and small soluble compounds and ions such as Cl^-, K^+, and Na^+ pass through the pores.

Other substances diffuse across the cell membrane without an expenditure of cellular energy, but these substances require the assistance of a carrier molecule. The carrier molecule is a protein and is located in the cell membrane. The carrier molecule binds to specific substances and assists their diffusion. This form of diffusion is referred to as facilitated diffusion.

Active transport does require an expenditure of cellular energy to convey substances. Hydrophilic (lipid-insoluble) substances are transported across the cell membrane by carrier molecules. The carrier molecule also is a protein and is specific with respect to the substances that they transport. For example, there are transport molecules for amino acids, ions, and sugars. Some toxic wastes enter bacterial cells on transport molecules that carry substrates and nutrients.

MOTILITY AND NUTRITION

Most bacteria move by the beating action of flagella. A flagellum is made of protein and enables the bacterium to travel up to $100\,\mu m/sec$. Younger bacteria tend to be more motile than older bacteria. Flagella are found on bacteria in three arrangements, monotrichous, lophotrichous, and peritrichous (Figure 3.6). Monotrichous and lophotrichous arrangements of flagella are found on one end of the bacterium. Only one flagellum is present in the monotrichous arrangement. In the peritrichous arrangement, many flagella are distributed over the entire surface of the bacterium.

Flagella are used to move (taxis) toward substrate (chemotaxis), light (phototaxis), or oxygen (aerotaxis). Each movement is positive taxis and is produced by concentration gradients for substrate and dissolved oxygen and intensity for light. Flagella also are used to move away from inhibitory and toxic wastes.

FACTORS AFFECTING BACTERIA GROWTH

The growth of bacteria in wastewater treatment plants and consequently treatment efficiency is influenced by a variety of nutritional factors and physical factors. Nutritional factors include the availability of substrates and nutrients. Physical factors include pH, temperature, and response to free molecular oxygen.

pH

Bacteria have an optimum pH at which they grow best. For most bacteria the optimum pH usually is near neutral (pH 7), and most bacteria do not grow at values ±1 unit of their optimum pH and cannot tolerate pH values below 4 or above 9.5. Most biological treatment units operate at pH values near neutral (6.8 to 7.2), and these units may experience operational problems at pH values below or above a near neutral pH value.

Operational problems that may occur in biological treatment units that experience pH values lower than 6.8 include the following:

- Decreased enzymatic activity
- Increase in hydrogen sulfide (H_2S) production
- Inhibition of nitrification
- Interruption of floc formation
- Undesired growth of filamentous fungi and some Nocardioforms

Operational problems that may occur in biological treatment units that experience pH values higher than 7.2 include the following:

- Decreased enzymatic activity
- Increase in ammonia (NH_3) production
- Inhibition of nitrification
- Interruption of floc formation

There are three groups of bacteria with respect to the conditions of acidity or alkalinity that they can tolerate. These groups include acidophiles, neutrophiles, and alkalinophiles. Acidophiles or acid-loving organisms grow at pH values lower than 5.4. *Thiobacillus* and *Sulfolobus* grow at pH values lower than 2, and many fungi prefer pH values lower than 5.

Neutrophiles grow at pH values from 5.4 to 8.5. Most bacteria in wastewater treatment plants are neutrophiles. Alkalinophiles or base-loving organisms grow at pH values from 7 to 11.5. The nitrifying bacteria, *Nitrosomonas* and *Nitrobacter*, are alkalinophiles.

In addition to the effect that pH has upon the activity of bacteria, there are two pH-related operational concerns. First, pH affects the degree of ionization of substrates, nutrients, and toxic wastes and their transportation into bacterial cells. Second, the use of substrates and production of wastes by bacteria may significantly change the pH of a biological treatment unit. Unless the pH of the unit is monitored and regulated, the change in pH may result in undesired bacterial activity and inefficient treatment of wastewater or sludge.

Examples of pH change in biological treatment units due to bacterial activity include the following:

- Denitrifying bacteria increase the pH of a biological treatment unit through the release of hydroxyl ions (OH^-).
- Fermentative bacteria decrease the pH of an anaerobic digester through the production of fatty acids.
- Methane-forming bacteria increase the pH of an anaerobic digester through use of fatty acids, especially acetate.
- Nitrifying bacteria decrease the pH of an aeration tank through the use and destruction of alkalinity.
- Organotrophic bacteria decrease the pH of a biological treatment unit through the production of carbonic acid (H_2CO_3) when they release carbon dioxide.

TEMPERATURE

Temperature exerts two significant effects upon a bacterial population. First, it affects the rate of diffusion of substrates and nutrients into bacterial cells. Second, it affects the rate of enzymatic activity. With increasing temperature the rate of diffusion of substrates and nutrients into bacteria cells increases, and the rate of enzymatic activity increases. The rates for diffusion and enzymatic activity decrease during decreasing temperatures.

Therefore, with increasing bacterial activity during warm wastewater temperatures, an operator of a wastewater treatment plant can decrease solids (bacteria) inventory and still maintain acceptable treatment of wastewater. However, with decreasing bacterial activity during cold wastewater temperatures, an operator of a wastewater treatment plant may need to increase solids inventory in order to maintain acceptable treatment of wastewater.

The impact of temperature upon bacterial activity is significant. For every 10°C rise in temperature, enzymatic activity nearly doubles. However, once the optimum temperature for enzymatic activity and cellular growth has been exceeded, enzymes become denatured (damaged) and can no longer efficiently catalyze biochemical reactions.

The temperature range of growth for an organism is determined by acceptable enzymatic activity in that range. The temperature range consists of three critical values. These values are (1) the minimum temperature that permits cell division, (2) the maximum temperature that permits cell division, and (3) the optimum temperature that permits the most rapid cell division.

There are three groups of bacteria with respect to the minimum and maximum temperatures that they remain active. These groups are psychrophiles, mesophiles, and thermophiles (Figure 3.7). Psychrophiles are cold-loving organisms and grow best at temperatures from 12°C to 18°C. Their range of temperatures for growth is from −10°C to 30°C. Psychrophiles are able to grow at low temperatures, because

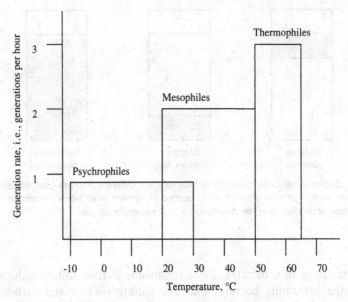

FIGURE 3.7 *Temperature ranges of growth and generation rates for psychrophiles, mesophiles, and thermophiles.*

their cell membrane contains a large quantity of unsaturated fatty acids. These acids help to maintain membrane fluidity at cold temperatures.

Mesophiles consist of the largest group of organisms and grow best at temperatures from 25°C to 40°C. Their range of temperatures for growth is 20°C to 50°C. Mesophiles are common inhabitants of the gastrointestinal tract of humans (body temperature approximately 37°C) and enter wastewater treatment plants in large numbers in human feces. They are present in very large numbers in the activated sludge process and the mesophilic anaerobic digester.

Thermophiles are heat-loving organisms and grow best at temperatures from 50°C to 65°C. Their range of temperatures for growth is from 35 to 75°C. Thermophiles are common inhabitants of thermophilic anaerobic digesters and thermophilic composting operations. Thermophiles are able to grow at high temperatures, because their cell membrane contains a large quantity of saturated fatty acids. These acids help to maintain membrane fluidity at high temperatures.

RESPONSE TO FREE MOLECULAR OXYGEN

Bacteria grow in the presence or absence of free molecular oxygen and can be placed in three groups according to their need for or response to free molecular oxygen. These groups are aerobes, anaerobes, and facultative anaerobes (Figure 3.8).

Aerobes require oxygen for the degradation of substrate. Examples of aerobic bacteria in activated sludge process include the filamentous organisms *Haliscomenobacter hydrossis* and *Sphaerotilus natans*, the floc former *Zoogloea ramigera*, and the nitrifying bacteria *Nitrosomonas* and *Nitrobacter*. Active aerobes are not found in anaerobic digesters. Anaerobic bacteria do not use free molecular oxygen

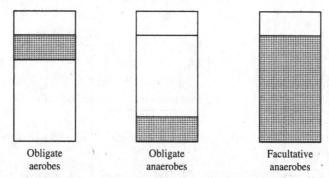

| Obligate | Obligate | Facultative |
| aerobes | anaerobes | anaerobes |

FIGURE 3.8 *Bacterial response to free molecular oxygen. Different bacteria (obligate aerobes, obligate anaerobes, and facultative anaerobes) incubated in nutrient broth tubes accumulate in different zones depending upon their need or sensitivity to free molecular oxygen.*

for the degradation of substrates. These organisms include sulfate-reducing bacteria and methane-forming bacteria that use sulfate (SO_4^{2-}) and carbon dioxide, respectively.

Between the aerobes and anaerobes are the facultative anaerobes. The term "facultative" implies the ability to live under different conditions. Facultative anaerobic bacteria have the ability to use free molecular oxygen or another molecule to degrade substrate. Facultative anaerobic bacteria have the most complex enzyme systems with respect to degradation of substrate. They have one enzyme system for the use of free molecular oxygen and have another enzyme system for the use of an alternate molecule to degrade substrate when oxygen is not available. However, they prefer oxygen to other molecules such as nitrate (NO_3^-) to degrade substrate, because they produce more offspring (sludge) from the same quantity of substrate with the use of free molecular oxygen than another molecule. Denitrifying bacteria including *Bacillus*, *Escherichia*, and *Pseudomonas* are facultative anaerobic bacteria.

There are two important groups of anaerobic bacteria. These groups are the oxygen-tolerant (aerotolerant) anaerobic bacteria and the oxygen-intolerant (obligate) anaerobic bacteria. The oxygen tolerant anaerobes can survive in the presence of free molecular oxygen. Oxygen tolerant anaerobes may or may not be active in the presence of free molecular oxygen. Obligate anaerobes such as the methane-forming bacteria die in the presence of free molecular oxygen.

Obligate anaerobes are not killed by free molecular oxygen. They are killed by superoxide (O_2^-) and hydrogen peroxide (H_2O_2). These products are formed when oxygen enters the bacterial cell. When oxygen enters the bacterial cell, it is converted to superoxide. Although superoxide is highly toxic, it is converted by the enzyme superoxide dismutase to oxygen and hydrogen peroxide (Equation 3.1). Although hydrogen peroxide also is highly toxic, it is converted by the enzyme catalase to oxygen and water (Equation 3.2). Most obligate anaerobic bacteria including the methane-forming bacteria lack the enzymes superoxide dismutase and catalase and succumb to the toxic effects of superoxide and hydrogen peroxide.

$$2O_2^- + 2H^+ \xrightarrow{\text{superoxide dismutase, if present}} O_2 + H_2O_2 \qquad (3.1)$$

TABLE 3.3   Concentration of Oxygen Needed in the Activated Sludge Process for Acceptable Biological Activity

Aerobic Biological Activity	Minimum Concentration (mg/liter)
Endogenous respiration	0.8
Floc formation	1
Nitrification	2–3
Control growth of low DO filaments	Correlated with chemical oxygen demand (COD) removed across aeration tank

$$2H_2O_2 \xrightarrow{\text{catalase, if present}} 2H_2O + O_2 \tag{3.2}$$

With respect to the quantity of oxygen necessary in activated sludge process to ensure acceptable biological activity by aerobe and facultative anaerobes, there are four activities of concern (Table 3.3).

4

Bacterial Groups

In order to reproduce all bacteria require (1) a source of carbon for the synthesis of new cellular materials, (2) a source of energy for cellular activity, and (3) inorganic nutrients. Growth factors such as amino acids and vitamins also may be required for reproduction.

Bacteria can be classified according to their similarities. Most often bacteria are placed largely into groups according to their carbon and energy sources. However, bacteria also are classified according to their structure (morphology), metabolism (physiology), and response to specific differential stains such as the Gram stain. Although bacteria can be classified in many ways, the most important classification of bacteria for wastewater treatment plant operators may be according to the roles they perform.

Bacteria commonly are classified according to their "trophic" nature. The two basic and major trophic needs for bacteria are carbon and energy (Table 4.1). Carbon is obtained primarily from two sources, while energy is obtained from three sources. The classification of bacteria based on carbon and energy sources also are used to categorize bacteria in wastewater treatment plants.

CLASSIFICATION OF BACTERIA ACCORDING TO CARBON AND ENERGY SOURCES THEY USE

The two major sources of carbon for bacteria are (1) organic compounds and (2) inorganic carbon and carbon dioxide. Bacteria that use organic compounds as their carbon source are called heterotrophs or organotrophs. Heterotrophs and organotrophs differ only with respect to their sources of hydrogen. However, most heterotrophs and organotrophs use common hydrogen sources, and the terms

TABLE 4.1 Classification of Bacteria According to Their
Carbon and Energy Sources

Carbon sources
    Inorganic compounds: autotrophs
    Organic compounds: heterotrophs or organotrophs
Energy sources
    Inorganic compounds: lithotrophs
    Organic compounds: heterotrophs or organotrophs
    Sunlight radiation: phototrophs

TABLE 4.2 Significant Lithotrophs in Wastewater Treatment Plants

Group	Genus	Energy Source	Waste Produced
Nitrifying bacteria	*Nitrosomonas*	NH_4^+	NO_2^-
	Nitrobacter	NO_2^-	NO_3^-
Sulfur bacteria	*Thiothrix*	H_2S	S^0
	Thiobacillus	S^0	SO_4^{2-}
Iron bacteria	*Siderocapsa*	Fe^{2+}	Fe^{3+}

heterotrophs and organotrophs are used interchangeably. Bacteria that obtain their electrons or hydrogen atoms (each hydrogen atom has one electron) from organic compounds are organotrophs. Heterotrophs or organotrophs degrade the cBOD in wastewater treatment plants. Bacteria that use carbon dioxide or another inorganic carbon compound, such as bicarbonate (HCO_3^-) and carbonate (CO_3^{2-}), as their carbon source are called autotrophs.

The three major sources of energy for bacteria are chemical oxidation reactions and sunlight. Bacteria that obtain their energy from chemical oxidation reactions are called chemotrophs. There are two types of chemotrophs, those that use organic compounds (heterotrophs) and those that use reduced inorganic compounds or elements. Bacteria that use reduced inorganic compounds or elements such as ionized ammonia (NH_4^+), hydrogen sulfide (H_2S), Fe^{2+}, and nitrite (NO_2^-) for energy are called lithotrophs. The term "litho" means rock or mineral. There are several lithotrophs that perform significant roles in wastewater treatment plants (Table 4.2). The most important lithotrophs are the nitrifying bacteria that degrade the nBOD; that is, they oxidize ionized ammonia and nitrite ions. Bacteria that obtain their energy from sunlight are called phototrophs.

CLASSIFICATION OF BACTERIA ACCORDING TO THE ROLES THEY PERFORM

Bacteria at wastewater treatment plants can be classified according to the roles that they perform (Table 4.3). Some bacteria perform positive roles in the treatment of wastewater, while other bacteria perform negative roles that contribute to inefficient treatment of wastewater, increased operational costs, and permit violations. Many bacteria perform positive and negative roles depending upon the operational conditions.

TABLE 4.3 Significant Groups of Wastewater Bacteria

Acetongenic bacteria	Hydrolytic bacteria
Coliforms	Methane-forming bacteria
Cyanobacteria	Nitrifying bacteria
Denitrifying bacteria	Nocardioforms
Fecal coliforms	Pathogenic bacteria
Fermentative (acid-forming) bacteria	Poly-P bacteria
Filamentous bacteria	Saprophytic bacteria
Floc-forming bacteria	Sheathed bacteria
Gliding bacteria	Spirochetes
Gram-negative aerobic cocci and rods	Sulfur-oxidizing bacteria
Gram-negative facultative anaerobic rods	Sulfur-reducing bacteria

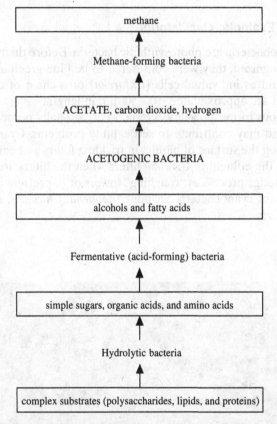

FIGURE 4.1 *Role of acetogenic (acetate-forming) bacteria in an anaerobic digester. Acetogenic bacteria perform a critical role in the anaerobic digester. These fermentative bacteria convert alcohols and fatty acids to acetate, the primary substrate for methane-forming bacteria or methanogenesis.*

Acetogenic Bacteria (Example: *Acetobacter*)

Members of the Acetobacteracae family produce acetate (CH_3COOH) and are important in the degradation of soluble cBOD to methane in anaerobic digesters. They are a special group of fermentative bacteria and convert organic acids, alcohols, and ketones to acetate, carbon dioxide, and hydrogen (Figure 4.1). Acetate is the principle substrate used by methane-forming bacteria for the production of methane (CH_4). Important acetogenic bacteria include *Acetobacter*, *Syntrobacter*, and *Syntrophomonas*.

Coliforms (Example: *Escherichia*)

Members of the coliform group of bacteria are Gram-negative rods that ferment the sugar lactose at 37°C and produce gas. The total coliform group includes these genera of the Enterobacteriacease family: *Citrobacter*, *Enterobacter*, *Escherichia*, *Hafnia*, *Klebisella*, *Serratia*, and *Yersinai*. The presence of coliform bacteria is an indicator of fecal contamination. *Escherichia* is the most representative genus of fecal contamination.

Cyanobacteria (Example: *Oscillatoria*)

Members of cyanobacteria are photosynthetic bacteria. Before their procaryotic cell structure was recognized, they were considered to be blue-green algae. Cyanobacteria may be found as individual cells (*Chlorella*) or a chain of cells or filament (*Oscillatoria*) that are approximately 100–500 μm in length.

Some filamentous forms of cyanobacteria do occasionally occur in the activated sludge process and may contribute to settleability problems. Cyanobacteria commonly are found on the surface of biofilm in trickling filters and enter the activated sludge process in the effluent of trickling filters when the filters are used upstream of the activated sludge process as "roughing" towers or to pretreat industrial wastewaters. Examples of cyanobacteria include *Anabaena*, *Chlorella*, *Euglena* (Figure 4.2), and *Oscillatoria*.

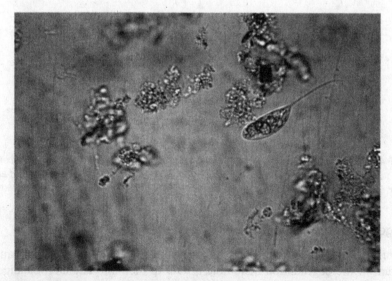

FIGURE 4.2 *Euglena. Euglena is considered to be not only a blue-green algae or cyanobacterium but also a pigmented flagellated protozoa.*

Denitrifying Bacteria (Example: *Bacillus*)

Denitrifying bacteria are facultative anaerobic bacteria that use nitrate (NO_3^-) in the absence of free molecular oxygen to degrade soluble cBOD. The use of nitrate results in the return of nitrogen to the atmosphere as molecular nitrogen (N_2) and nitrous oxide (N_2O). Denitrification is used to satisfy a total nitrogen discharge requirement for a wastewater treatment plant. Denitrification also causes "clumping" in secondary clarifiers and foaming in anaerobic digesters. Although there are numerous genera that contain denitrifying bacteria, the three genera that contain the most species of denitrifying bacteria are *Alcaligenes*, *Bacillus*, and *Pseudomonas*.

Fecal Coliforms (Example: *Escherichia*)

Fecal coliforms include all coliforms that are exclusively fecal in origin and can ferment lactose or produce colonies at 45°C. Fecal coliform is used as an indicator of fecal contamination and consists principally of species of the genus *Escherichia*.

Fermentative (Acid-Forming) Bacteria (Example: *Proteus*)

Fermentative or acid-forming (acidogenic) bacteria convert amino acids, fatty acids, and sugars to organic acids including acetate (CH_3COOH), butyrate ($CH_3CH_2CH_2COOH$), formate ($HCOOH$), lactate ($CH_3CHOHCOOH$), and propionate (CH_3CH_2COOH). Fermentative bacteria are important in anaerobic digesters where they convert complex substrates to simple substrates that can be used by methane-forming bacteria (Figure 4.1). Fermentative bacteria also are important in biological phosphorus removal units where they produce the necessary organic acids that are required for the uptake of phosphorus by poly-P bacteria. There are numerous genera of fermentative bacteria and include *Bacteroides*, *Bifidobacteria*, *Clostridium*, *Escherichia*, *Lactobacillus*, and *Proteus*.

Filamentous Bacteria (Example: *Haliscomenobacter*)

There are approximately 30 filamentous organisms that contribute to settleability problems in activated sludge processes due to their rapid and undesired growth. Of these filamentous organisms, 10 bacteria are responsible for most bulking episodes. Although better known for operational problems, filamentous bacteria due contribute to the degradation of soluble cBOD and floc formation. The most commonly occurring filamentous bacteria are *Haliscomenobacter hydrossis*, *Microthrix parvicella*, Nocardioforms, *Sphaerotilus natans*, *Thiothrix*, type 0041, type 0092, type 0675, type 1701, and type 021N.

Floc-Forming Bacteria (Example: *Zoogloea*)

Floc-forming bacteria initiate floc formation in the activated sludge process. With increasing sludge age, floc-forming bacteria produce the necessary cellular components needed to stick together or agglutinate. Only a small number of bacteria are floc formers and include *Achromobacter*, *Aerobacter*, *Citromonas*, *Flavobacterium*, *Pseudomonas*, and *Zoogloea*.

Gliding Bacteria (Example: *Beggiatoa*)

There are three motile or gliding filamentous organisms in the activated sludge process that contribute to settleability problems. These organisms are *Beggiatoa*, *Flexibacter*, and *Thiothrix*.

Gram-Negative, Aerobic Cocci and Rods (Example: *Acetobacter*)

Gram-negative, aerobic cocci and rods make up approximately 20% of the bacteria in the activated sludge process. They are involved in biological phosphorus removal, degradation of soluble cBOD, floc formation, and nitrification. Genera of bacteria that contain Gram-negative, aerobic cocci and rods include *Acetobacter*, *Acinetobacter*, *Alcaligens*, *Nitrobacter*, *Nitrosomonas*, *Pseudomonas*, and *Zoogloea*.

Gram-Negative, Facultative Anaerobic Rods (Example: *Escherichia*)

Gram-negative, facultative anaerobic rods make up approximately 80% of the bacteria in the activated sludge process and a significant number of the bacteria in the anaerobic digester. They are involved with biological phosphorus removal, acetate production, acid production, degradation of soluble cBOD, denitrification, floc formation, and hydrolysis of cBOD. Genera of bacteria that contain Gram-negative, facultative anaerobic rods include *Aeromonas*, *Escherichia*, *Flavobacterium*, *Klebsiella*, *Proteus*, and *Salmonella*.

Hydrolytic Bacteria (Example: *Bacteriodes*)

Numerous facultative anaerobic bacteria and anaerobic bacteria make up the hydrolytic bacteria. The more important hydrolytic bacteria are the anaerobic bacteria. The hydrolytic bacteria produce exoenzymes that solubilize complex insoluble substrates into simple soluble substrates that can be absorbed and degraded by an even larger number of bacteria. Acceptable enzymatic activity by hydrolytic bacteria is essential for the proper operation of suspended growth, municipal anaerobic digesters that must degrade large quantities of colloidal substrates and particulate substrates. The most important hydrolytic bacteria are *Bacteroides*, *Bifidobacteria*, and *Clostridium*.

Methane-Forming Bacteria (Example: *Methanobacterium*)

Methane-forming bacteria or methanogens produce methane (CH_4) from a limited number of substrates in anaerobic digesters. Methane production occurs through two major routes: the splitting of acetate and the use of carbon dioxide and hydrogen. Examples of genera of methane-forming bacteria include *Methanobacterium*, *Methanococcus*, *Methanomonas*, and *Methanosarcinia*.

Nitrifying Bacteria (Example: *Nitrosomonas*)

Nitrifying bacteria are strict aerobes. They oxidize ionized ammonia (NH_4^+) to nitrite (NO_2^-) and oxidize nitrite to nitrate (NO_3^-). Bacteria that oxidize ionized ammonia include *Nitrosomonas* and *Nitrosospira*. Bacteria that oxidize nitrite include *Nitrobacter* and *Nitrospira*.

Nocardioforms (Example: *Nocardia*)

Nocardioforms or actinomycetes are a specialized group of Gram-positive and spore-forming bacteria. These filamentous organisms are relative short (<50 μm) and highly branched. Nocardioforms display some growth characteristics such as true branching that are found in fungi. *Nocardia* and related genera that are most often associated with foam production in the activated sludge process include *Actinomadura*, *Arthrobacter*, *Corynebacterium*, and *Micromonospora*. *Nocardia* is the most commonly observed foam-producing actinomycete, and the species of *Nocardia* most often reported as problematic include *N. amarae*, *N. asteroides*, *N. caviae*, *N. pinesis*, and *N. rhodochrus*.

Pathogenic Bacteria (Example: *Camplyobacter*)

Numerous pathogenic (disease-causing) bacteria enter wastewater treatment plants from domestic wastewater, slaughterhouse wastewater, and inflow and infiltration (I/I). There are two types of pathogenic bacteria: "true" pathogens and "opportunistic" pathogens. The bacterial pathogens that represent significant risks of disease transmission to wastewater personnel are *Camplyobacter jejuni* and *Leptospira interrogans*.

Poly-P Bacteria (Example: *Acinobacter*)

Poly-P bacteria or phosphorus accumulating organisms (PAO) are used in biological phosphorus removal units. By recycling the bacteria through anaerobic and aerobic zones, the bacteria remove orthophosphate ($H_2PO_4^-$/HPO_4^{2-}) from the wastewater in quantities greater than normal cellular needs. Genera of bacteria that contain Poly-P bacteria include *Acinobacter*, *Aerobacter*, *Beggiatoa*, *Enterobacter*, *Klebsiella*, and *Proteobacter*.

Saprophytic Bacteria (Example: *Micrococcus*)

Saprophytic bacteria feed upon dead organic matter (cBOD). Saprophytic bacteria are organotrophs, and many are floc-forming bacteria. The major saprophytic bacteria include *Achromobacter*, *Alcaligenes*, *Bacillus*, *Flavobacterium*, *Micrococcus*, and *Pseudomonas*.

Sheath Bacteria (Example: *Sphaerotilus*)

Sheath bacteria consist of a chain of Gram-negative cells that are surrounded by a transparent tube or sheath (Figure 4.3). When the cells leave the sheath, they become motile by means of flagella and are referred to as swamer cells. The swamer cells quickly produce sheaths and form filamentous chains. There are two sheathed, filamentous bacteria in the activated sludge process: *Haliscomenobacter hydrossis* and *Sphaerotilus natans*.

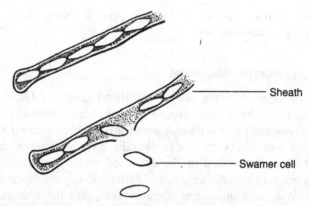

FIGURE 4.3 *Sheath filamentous organism. Some filamentous organisms such as* Sphaerotilus natans *possess a translucent sheath that surrounds the individual cells in the bacterial chain. When the sheath breaks, individual cells or swamer cells are released. These cells reproduce to form more filamentous organisms.*

Spirochetes (Example: *Spirochaeta*)

Spirochetes are helical-shaped, motile bacteria. There are three types of spirochetes: (1) aerobic, (2) facultative anaerobic, and (3) anaerobic. Proliferation of each type of spirochete occurs within wastewater with changes in oxygen tension. Free-living spirochetes are found in a variety of aquatic habitats, including wastewater. Pathogenic spirochetes such as *Leptospira* live in bodily fluids. *Spirochaeta* is a genus of free-living spirochetes that is found in wastewater.

Sulfur-Oxidizing Bacteria (Example: *Thiobacillus*)

Sulfur-oxidizing bacteria oxidize inorganic sulfur by adding oxygen to sulfur. Sulfur-oxidizing bacteria obtain energy from the oxidation of sulfur. There are non-filamentous and filamentous sulfur-oxidizing bacteria. Nonfilamentous bacteria include *Thiobacillus*, *Thiospirillopsis*, and *Thiovulum*. Filamentous bacteria include *Beggiatoa* and *Thiothrix*.

Sulfur-Reducing Bacteria (Example: *Desulfovibrio*)

Sulfur-reducing bacteria are anaerobes and use sulfate to degrade substrates. Principal sulfur-reducing bacteria are *Desulfovibrio* and *Desulfotomaculum*.

5

Bioaugmentation

Bioaugmentation or biomass enhancement is the addition of commercially prepared bacterial cultures to a wastewater treatment system to (1) increase the density of desired bacteria and their enzymes and (2) achieve a specific operational goal—for example, decrease sludge production or control malodor production. The addition of bacterial cultures increases the density of desired bacteria without significantly increasing the solids inventories and solids residence times of an activated sludge process or anaerobic digester. Sufficient addition of bioaugmentation products may enable an operator to decrease MLVSS concentration and MCRT. Decreases in MLVSS and MCRT help to control the undesired growth of filamentous organisms.

Treatment efficiency, permit compliance, and operational costs at a municipal wastewater treatment plant are influenced greatly by the enzymatic activities and abilities of a large population and diversity of coli-aerogens. Coli-aerogens are bacteria that inhabit the gastrointestinal tract of humans and enter wastewater treatment plant in fecal waste.

Examples of significant activities of coli-aerogens that are of important to wastewater treatment plants include:

- Nutrient and dissolved oxygen requirements
- Products obtained from the degradation of substrates
- Rates of degradation of the substrates
- Types of substrates that can be degraded

Examples of significant abilities of coli-aerogens that are of importance to wastewater treatment include

- Adverse conditions that are tolerated
- Competition with other organisms
- Floc-forming ability
- pH growth range
- Temperature growth range

The activities and abilities of coli-aerogens are supported by a smaller population of several important genera of saprophytic and nitrifying bacteria that enter the treatment plant as soil and water organisms through inflow and infiltration (I/I). The saprophytic bacteria and their enzyme systems are more efficient in degrading a larger variety of substrates than the coli-aerogens. Also, many saprophytic bacteria have unique abilities that enable them to survive and remain active under harsh environmental or operational conditions that are not tolerated well by the coli-aerogens.

Saprophytic bacteria are primarily responsible for the degradation of organic compounds (substrates) in nature. However, saprophytic bacteria do not enter wastewater treatment plants in significant numbers and do not grow in large numbers in wastewater or sludge, due to the presence of very large numbers of coli-aerogens that enter wastewater treatment plants. Therefore, the efficient enzymatic activities and unique abilities of the saprophytic bacteria are "diluted" by the coli-aerogens. Due to the relatively small number of saprophytic bacteria in a wastewater treatment plant as compared to the large number of coli-aerogens, a wastewater treatment plant may experience difficulties in treating specific substrates, tolerating adverse conditions, or correcting an operational problem.

The addition of bioaugmentation products is to improve or correct treatment plant performance. The products or bacterial cultures used at a treatment plant are selected according to (1) the needs of the treatment plant and (2) the activities and abilities of the bacterial cultures to address the operational problem.

Bacterial cultures may be added to several locations at a wastewater treatment system (Table 5.1). The location for the addition of the bacterial cultures is selected according to the needs of the treatment system and the adjustment period of the bacteria. The adjustment period is the time required by the bacteria to produce their enzymes in a new environment such as the aeration time or anaerobic digester. The longer the adjustment time that is required, the greater the distance (detention time) from the treatment unit the bacteria are added.

Although bioaugmentation products are introduced to a specific location for use in a specific tank, the bacteria are transferred throughout the treatment plant—that

TABLE 5.1 Commonly Used Addition Points for Bioaugmentation Products

Lift station
Conveyance system
Headworks
Primary clarifier influent
Primary clarifier effluent
Aeration tank influent

is, activated sludge, aerobic digester, and anaerobic digester. Bacterial cultures are added to a treatment unit at an introductory dose and maintenance dose. If an introductory dose is used, the dose usually is ≥2 ppm and may be applied for 2–4 weeks. In most cases the maintenance dose is approximately 2 ppm and may be applied daily, weekly, or as needed.

Bioaugmented saprophytic bacteria do reproduce in wastewater treatment plants, but they cannot outnumber the coli-aerogens that continuously enter the process in very large numbers. Coli-aerogens are present in millions per milliliter of mixed liquor and billions per gram of floc particle. Saprophytic bacteria are added to a level where the impact of their activities and abilities can be observed.

Bioaugmented saprophytic bacteria are not pathogenic (disease-causing). However, the some preservatives that are used to arrest the growth of the bacteria during storage and shipping may cause an allergic reaction with some individuals. Therefore, appropriate protective clothing such as long sleeve uniforms or shirts, dust masks, and splash shields or eye goggles should be used when handling bioaugmentation products. Individuals who come in contact with the products should wash or shower.

Selected saprophytic bacteria are obtained from soil and water samples from a variety of habitats. They are screened to determine their enzymatic activities and abilities. Screening determines

- Efficiency and rate of degradation of organic compounds (substrates)
- Operational conditions that are tolerated
- Variety of organic compounds (substrates) that can be degraded

Saprophytic bacteria are grown to a relatively large population on a carbon source. Their growth is arrested through suspension, freeze-drying, or air-drying techniques, and the bacteria are packaged in liquid or dry forms.

Although there are numerous genera and species of saprophytic bacteria that are used for specific and general purposes, there are several common genera that are used in many bioaugmentation products (Table 5.2). In addition to saprophytic bacteria, nitrifying bacteria may be added used (Table 5.3) as well as fungi.

TABLE 5.2 Commonly Used Genera of Saprophytic Bacteria in Bioaugmentation Products

Genus	Enzymatic Activity
Aerobacter	In anaerobic digester, it coverts complex organic compounds into volatile fatty acids that are converted to methane. It increases gas production and decreases sludge production in anaerobic digester.
Bacillus	It produces a large number of exoenzymes that solubilize and degrade a large variety of carbohydrates, lipids, and proteins. It decreases sludge production.
Cellulomonas	It solubilizes and degrades cell wall material such as cellulose within vegetative tissue. It decreases sludge production.
Pseudomonas	It degrades difficult biodegradable compounds such as phenols and phenolic compounds. It decreases sludge production and reduces toxic upsets caused by phenols and phenolic compounds.
Rhodopseudomonas	It degrades difficult biodegradable compounds. It decreases sludge production.

TABLE 5.3 Nitrifying Bacteria Used in Bioaugmentation Products

Genus	Enzymatic Activity
Nitrobacter	It converts nitrite (NO_2^-) to nitrate (NO_3^-), resulting in reduced nitrogenous loading to the receiving waters. It initiates or improves nitrification at low MLVSS concentration or cold temperature.
Nitrosomonas	It converts ionized ammonia (NH_4^+) to nitrite (NO_2^-), resulting in reduced nitrogenous loading to the receiving waters. It initiates or improves nitrification at low MLVSS concentration or cold temperature.

Saprophytic bacteria can efficiently degrade readily degradable cBOD and difficult-to-degrade cBOD. Degradation is achieved through the use of unique enzyme systems and the production and release of exoenzymes (Figure 5.1). Exoenzymes released by saprophytic bacteria diffuse through the floc particle and solubilize carbohydrates, lipids, and proteins that have been adsorbed to the floc particle. Once solubilized, the sugars (from carbohydrates), fatty acids (from lipids), and amino acids (from proteins) can be degraded further by not only the saprophytic bacteria but also the coli-aerogens. Also, many saprophytic bacteria can compete more effectively for available nutrients and dissolved oxygen in the bulk solution than many coli-aerogens and filamentous organisms. This effective competition for dissolved oxygen and nutrients helps to provide for acceptable wastewater treatment during marginal concentrations of dissolved oxygen and nutrients and may help to prevent the undesired growth of filamentous organisms.

An example of the ability of saprophytic bacteria to solubilize and degrade cBOD resulting in improved treatment plant performance (i.e., decreased sludge production) is the solubilization and degradation of cellulose (Figure 5.2). Cellulose is an insoluble polysaccharide or starch consisting of numerous units or mers (polymer) of glucose. The glucose units are held in an insoluble form due to the unique chemical bond between each glucose unit. Cellulose is found in plants and is not degraded by humans or the coli-aerogens in the gastrointestinal tract. Therefore, much cellulose is found in fecal waste. Cellulose is found in primary and secondary sludges at wastewater treatment plants. Cellulose makes up 8–15% of primary and secondary sludges.

Cellulomonas is a genus of bacteria having species that produce the exoenzyme cellulase. The enzyme is capable of breaking the unique chemical bonds between the glucose units in cellulose. However, *Cellulomonas* typically is not found in large numbers in the indigenous bacterial population of an activated sludge process, but *Cellulomonas* may be augmented to the process.

By adding *Cellulomonas* to the activated sludge process, cellulose is solubilized; that is, individual glucose units are separated from the starch and dissolve in the wastewater. Once in solution, glucose is absorbed and degraded by *Cellulomonas* and many other bacteria including coli-aerogens. The degradation of glucose results in the production of carbon dioxide, water, and new bacterial cells (sludge or MLVSS). Approximately 0.6 pound of cells or sludge and 0.4 pound of carbon dioxide and water are produced from each pound of glucose degraded. A pound of cellulose that is solubilized and degraded represents approximately 0.6 pound of cells or sludge wasted from the activated sludge process. A pound of cellulose that is not solubilized and degraded represents one pound of solids or sludge that is

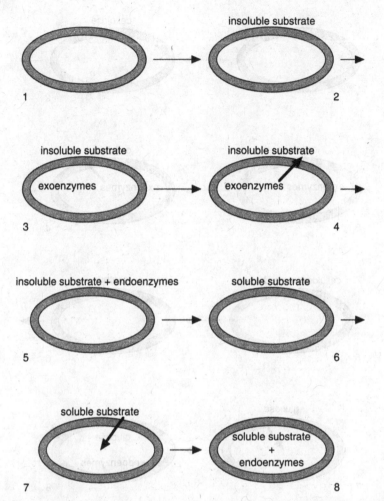

FIGURE 5.1 *Production and release of exoenzymes. When an insoluble substrate becomes adsorbed to the surface of an exoenzyme-producing bacterium (2), exoenzymes are produced within the cell (3). The exoenzymes are released by the bacterium (4) and "attack" the insoluble substrate. The attack upon the insoluble substrate results in the solublization of the substrate (5) and the production of soluble substrate (6). The soluble substrate then is absorbed by the exoenzyme-producing bacterium (7) and finally degraded by endoenzymes within the exoenzyme-producing bacterium.*

wasted from the activated sludge process. The use of *Cellulomonas* to solubilize cellulose represents a reduction of approximately 40% in sludge production for cellulose. *Cellulomonas* added to the activated sludge process is transferred to the aerobic or anaerobic digester in secondary sludge. *Cellulomonas* continues to degrade cellulose in the aerobic or anaerobic digester.

Bioaugmentation products may be used to correct several operational problems or improve treatment efficiency (Table 5.4). Some applications of bioaugmentation products (e.g., decreased sludge production) may result in monetary savings to the wastewater treatment plant. Sulfur-containing malodorous compounds also can be degraded with appropriate bacterial cultures. For example, bacteria in the genera

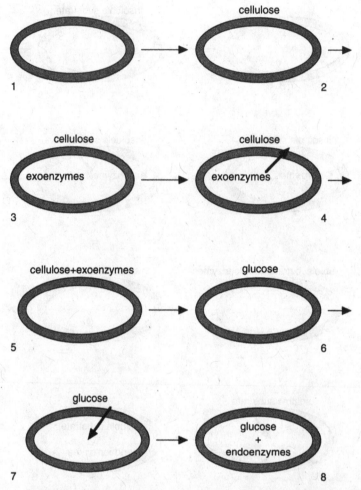

FIGURE 5.2 *Degradation of cellulose. Cellulose is an insoluble starch molecule that consists of many units of glucose bonded together in chain-like fashion. When cellulose comes in contact with the bacterium Cellulomonas, cellulose is adsorbed to the surface of the bacterium (2). Once cellulose is absorbed, Cellulomonas produces the exoenzyme cellulase (3) that is released to the surface of the bacterium and "attacks" cellulose (4). Cellulase solublizies cellulose (5), and glucose is produced from the attack by cellulose (6). Soluble glucose is then absorbed by Cellulomonas (7) and is degraded by endoenzymes (8).*

Thiobacillus and *Hyphomicrobium* can degrade methyl sulfide, dimethyl sulfide, and dimethyl disulfide.

Before adding bioaugmentation products, the following information should be reviewed in order to select and apply the appropriate bacterial cultures:

• Evaluate the treatment plant's existing conditions.
• Identify the specific conditions to be addressed with bioaugmentation products.

TABLE 5.4 Applications for Bioaugmentation Products

Application	Example
Control of foam production	Addition of bacterial cultures that produce enzymes that degrade lipids (lipase) or surfactants
Control of malodors	Addition of bacterial cultures (*Pseudomonas*) that degrade organic sulfur compounds before they are released to the atmosphere
Nitrification, cold weather	Addition of *Nitrosomonas* and *Nitrobacter* to the aeration tank to initiate or improve cold weather nitrification or addition of saprophytic bacteria to remove cBOD more quickly to prove longer time period for nitrification to occur
Reduce filamentous growth	Addition of sufficient saprophytic bacteria to reduce MLVSS concentration and MCRT
Reduce sludge production	Addition of saprophytic bacteria to solubilize particulate and colloidal cBOD
Increase digester gas production	Addition of facultative anaerobic and anaerobic bacteria to hydrolyze particulate and colloidal BOD and produce substrate (fatty acids) for methane-forming bacteria
Improve BOD and SS removal efficiency	Addition of saprophytic bacteria that are more efficient and tolerant of existing operational condition
Provide resistance to some forms of toxicity	Addition of saprophytic bacteria that can safely bioaccumulate heavy metals in slime and cell walls or can degrade organic toxicants

- Collect and review appropriate data for use in selecting appropriate bioaugmentation products.
- Review possible effects of bioaugmentation products upon indigenous biomass.
- Review storage and application procedures, including activation of bacterial cultures.
- Determine time period needed to observed the impact of bioaugmentation products.
- Determine tests or observations that indicate that the bioaugmentation products are working.
- Determine the costs for bioaugmentation products and any anticipated monetary savings.

6

Pathogenic Bacteria

A large number and diversity of pathogenic (disease-causing) bacteria enter sanitary sewer systems and wastewater treatment plants on a daily basis. Pathogenic bacteria enter sewer systems from (1) domestic wastewater, (2) industrial wastewaters such as slaughterhouses, (3) cat and dog excrement through inflow and infiltration (I/I), and (4) rats that inhabit the sewer system. Pathogenic bacteria can be found in the sewer system in the wastewater, sediment, and biofilm and at wastewater treatment plants in wastewater, sludges, bioaerosols, contaminated surfaces, foam, recycle streams, and scum.

There are several pathogenic bacteria of concern to wastewater personnel (Table 6.1). Of these pathogens, *Camplyobacteria jejuni* and *Leptospira interrogans* represent an elevated risk of disease transmission to wastewater personnel.

Camplyobacter jejuni is a Gram-negative, rod-shaped bacterium that causes gastroenteritis. Symptoms of gastroenteritis include abdominal pain, fever, and bloody diarrhea. *Camplyobacter jejuni* is transmitted through fecal waste.

Leptospira interrogans is a Gram-negative spirochete that causes leptospirosis, an infection of the kidneys. Symptoms of leptospirosis include fever, headache, jaundice, and kidney and liver damage. *Leptospira interrogans* is transmitted through urine and is carried by rats.

There are two types of pathogenic bacteria: "true" pathogens and "opportunistic" pathogens. True pathogens such as *Shigella* are aggressive and are transmitted from person to person and by contact with animals and their wastes. Opportunistic pathogens such as *Escherichia coli* are typically found on or in the human body and do not cause disease unless the body's immune system is weakened by injury, a true pathogen, or physiological disease.

Some pathogenic bacteria produce endospores or capsules. Endospores and capsules protect pathogenic bacteria from harsh environmental conditions and

49

TABLE 6.1 Pathogenic Bacteria Associated with Wastewater

Bacterium/Bacteria	Disease
Actinomyces israelii	Actinomycosis
Camplyobacter jejuni	Gastroenteritis
Clostridium perfringens	Gangrene (gas gangrene)
Clostridium tetani	Tetanus
Escherichia coli—enteroinvasive	Gastroenteritis
Escherichia coli—enteropathogenic	Gastroenteritis
Escherichia coli—enterotoxigenic	Gastroenteritis
Escherichia coli—enterohemorrhagic 0157:H7	Gastroenteritis and hemolytic uremic syndrome
Leptospira interrogans	Leptospirosis
Mycobacterium tuberculosis	Tuberculosis
Nocardia spp.	Nocardosis
Salmonella paratyphi	Paratyphoid fever
Salmonella spp.	Salmonellosis
Salmonella typhi	Typhoid fever
Shigella spp.	Shigellosis
Vibrio cholerae	Cholera (Asiatic chlorea)
Vibrio parahaemolyticus	Gastroenteritis
Yersinia enterocolitica	Yersiniosis (bloody diarrhea)

disinfection. Examples of endospore-forming or spore-forming bacteria include *Bacillus* and *Clostridium*. Examples of capsule-forming bacteria include *Bacillus* and *Streptococcus*.

Endotoxins and exotoxins produced by pathogenic bacteria also represent a concern to wastewater personnel. Endotoxins are components of the bacterial cell wall and are released by bacteria when they die. Endotoxins diffuse into the host's tissues and cause nonspecific or localized reactions in individuals such as gastrointestinal or respiratory diseases.

Exotoxins are found inside the bacterial cell and are excreted by living cells into their surrounding medium and are absorbed by the host's tissues. Exotoxins are very potent toxins and highly specific with respect to the reactions they cause—for example, neurotoxins and cardiac muscle toxins. Examples of disease caused by exotoxins include botulism and tetanus.

Pathogenic bacteria enter wastewater treatment plants as suspended bacteria, associated with solids and associated with host cells. Pathogenic bacteria are removed in the primary clarifier as bacteria associated with solids. When the solids settle in the clarifier, the bacteria are removed from the waste stream. Suspended and cell-associated pathogenic bacteria are removed in the activated sludge process when they are adsorbed to floc particles. Although many pathogenic bacteria are destroyed in the harsh environment of the wastewater and the activated sludge process, many pathogenic bacteria are transferred in primary and secondary sludges to the anaerobic digester. In properly operating anaerobic digesters, pathogenic bacteria are destroyed in large numbers due to the following conditions:

- Lack of nutrients due to competition with other bacteria
- Long solids retention time (SRT)
- Septicity

Risks associated with disease transmission to wastewater personnel can be reduced using common sense, proper hygiene measures, and appropriate protective equipment. These practices deny portals of entry (ingestion, inhalation, and invasion) to pathogenic bacteria. A comprehensive review of pathogens is provided in *Wastewater Pathogens* in the Wastewater Microbiology Series.

Part II

Enzymes and
Bacterial Growth

7

Enzymes

Enzymes are a special group of complex proteins found in all living organisms. Most cells contain hundreds of enzymes and are continuously synthesizing enzymes. Enzymes act as catalysts of biochemical reactions. Enzymes accelerate the rate reactions as much as 1,000,000 times the rate of uncatalyzed reactions and permit the occurrence of biochemical reactions at temperatures that living cells can tolerate.

Some enzymes contain nonprotein groups or cofactors. Cofactors include coenzymes or organic molecules such as biotin (Figure 7.1) and metallic activators or inorganic ions such as cobalt (Co^{2+}), copper (Cu^{2+}), magnesium (Mg^{2+}), potassium (K^+), and zinc (Zn^{2+}). Cofactors improve the efficiency and rate of enzymatic activity. Because enzymes are not consumed and do not undergo structural damage during biochemical reactions, they can be used repeatedly.

Enzymes provide a surface or active site on which biochemical reactions can occur. The active site is a binding site; that is, it forms a weak bond with its substrates, the molecules that the enzyme acts upon. However, enzymes are specific with respect to the substrates that they can degrade or compounds they can synthesize.

When a substrate comes in contact with an appropriate site of an enzyme, an enzyme–substrate complex is formed. Once the complex is formed, chemical bonds in the substrate are weakened, and the substrate may be degraded (catabolism) to simpler molecules (Figure 7.2) or assimilated (anabolism) to more complex molecules (Figure 7.3). Catabolic reactions result in a decrease in sludge production— for example, the degradation of stored food in bacterial cells. Anabolic reactions result in an increase in sludge production—for example, the transformation and assimilation of sugars into new bacterial cells or sludge.

Enzymes usually have a high degree of specificity; that is, they are very specific with respect to (a) the substrates that they can degrade and (b) the compounds and

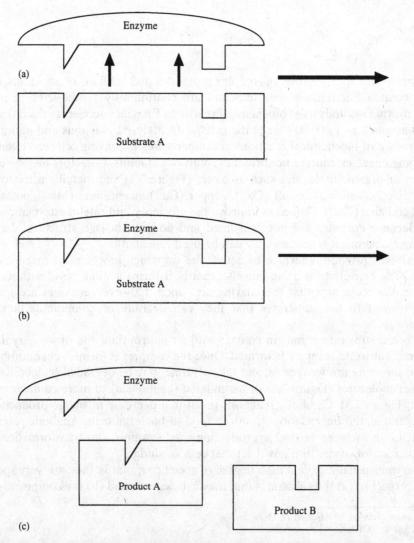

FIGURE 7.1 *Biotin. Biotin is a water soluble, organic vitamin B complex that is required for bacterial growth.*

FIGURE 7.2 *Catabolism. During catabolism, a large substrate molecule (A) bonds to an enzyme (enzyme–substrate complex) according to the shapes and charges of the substrate and enzyme (b). Once bonded, the enzyme degrades or breaks the substrate (c). Catabolism results in the production of smaller substrates and the release of energy.*

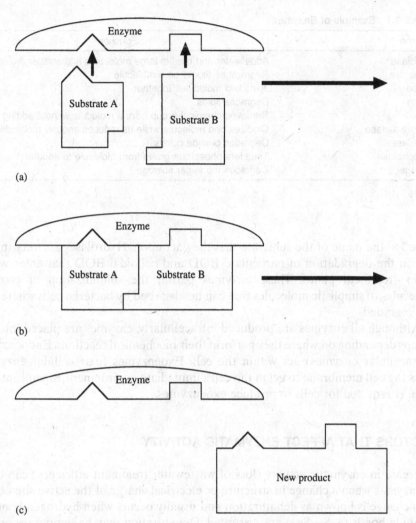

(a)

(b)

(c)

FIGURE 7.3 *Anabolism. During anabolism, small substrate molecules (A and B) bond to an enzyme (enzyme–substrate complex) according to the shapes and charges of the substrates and enzyme (b). Once bonded, the enzyme joins the substrates together to form a larger substrate. Anabolism results in the production of new products such as those used in the synthesis of cellular material.*

ions that they can assimilate. For example, some enzymes can degrade a large number of carbohydrates, while other enzymes can degrade only a small number of carbohydrates or a specific carbohydrate.

The specificity of the biochemical reactions that an enzyme may catalyze is due to the shape and electrical charge of the active site of the enzyme. The shape of an enzyme is produced through the bonding of thiol groups (—SH), while its electrical charge is due to the ionization of hydrogen bonds. If an enzyme is capable of acting on more than one substrate, it usually acts on substrates with the same functional group [e.g., carboxyl (—COOH) or hydroxyl (—OH)], or the same kind of chemical bond (Table 7.1). For example, proteolytic (protein-splitting) enzymes break peptide bonds in proteins. Enzymes are usually named by adding the suffix

TABLE 7.1 Example of Enzymes

Enzyme	Function
Hydrolase	Adds water and breaks large molecules into smaller molecules
Isomerase	Rearranges atoms on a molecule
Ligase	Joins two molecules together
Lipase	Degrades lipids
Lyase	Removes chemical group from a molecule without adding water
Oxidoreductase	Oxidizes one molecule while its reduces another molecule
Peptidase	Degrades peptide bonds
Phosphatase	Transfers phosphate group from molecule to another
Sucrase	Degrades the sugar sucrose

"-ase" to the name of the substrate that they act upon. Hydrolases are very important in the degradation of particulate BOD and colloidal BOD that enter wastewater treatment plants. These enzymes permit the solubilization of complex molecules to simplistic molecules that can be absorbed by bacterial cells where they are degraded.

Although all enzymes are produced intracellularly, enzymes are placed into two groups depending on where they perform their biochemical reactions. Endoenzymes (intracellular enzymes) act within the cell. Exoenzymes (extracellular enzymes) cross the cell membrane to act in the cell's immediate environment. Significant time often is required for cells to produce exoenzymes.

FACTORS THAT AFFECT ENZYMATIC ACTIVITY

Decrease in enzymatic activity (loss of wastewater treatment efficiency) can occur in enzymes when a change in structure or electrical charge of the active site occurs. This change is known as denaturation and usually occurs when hydrogen bonds or disulfide bonds (—S—S—) are disrupted. Denaturation may be temporary or permanent. Denaturation may occur in the presence of elevated temperatures, depressed or elevated pH values, and inhibitory or toxic wastes including heavy metals and oxidizing agents such as chlorine, hydrogen peroxide (H_2O_2), and potassium permanganate ($KMnO_4$).

8

Hydrolytic Bacteria

A large and diverse population of bacteria and their enzymes are necessary to degrade the large quantity and variety of substrates that enter a biological treatment unit. Because different groups of bacteria reproduce at different rates, the mean cell residence time (MCRT) or solids retention time (SRT) of a treatment unit must be adjusted to grow the required bacterial population and their appropriate enzymes for appropriate biological activities. These activities include (1) the degradation of specific substrates, (2) floc formation, and (3) the removal of specific pollutants such as phosphorus (Tables 8.1 and 8.2). The MCRT also can be adjusted to prevent the degradation of a specific substrate. For example, the MCRT of an activated sludge process can be adjusted to promote biological phosphorus removal and prevent nitrification (Table 8.1).

The MCRT or SRT necessary for appropriate biological activity is influenced by temperature. With increasing temperature the activity of the bacteria or biomass in a biological treatment unit increases; therefore a smaller number of bacteria or quantity of biomass is required. This permits a decrease in MCRT or SRT. With decreasing temperature the activity of the bacteria or biomass in a biological treatment unit decreases, and therefore a larger number of bacteria or quantity of biomass is required. This requires an increase in MCRT or SRT. Bacteria in a biological treatment unit may be present mostly in the dispersed state (Figure 8.1) such as a suspended growth anaerobic digester or the flocculated state (Figure 8.2) such as the activated sludge process.

In biological treatment units, substrates are absorbed or adsorbed by bacteria (Figure 8.3). Absorbed substrates are those that are soluble, are simple in structure, and quickly enter bacterial cells. Examples of absorbed substrates or soluble cBOD include acetate (CH_3COOH), ethanol (CH_3CH_2OH), and glucose ($C_6H_{12}O_6$). These substrates are quickly degraded by endoenzymes; as a result, the substrates present

Wastewater Bacteria, by Michael H. Gerardi
Copyright © 2006 John Wiley & Sons, Inc.

TABLE 8.1 MCRT (Approximate Days) Required for the Initiation of Significant Biological Activities in the Activated Sludge Process

Biological Activity	Approximate MCRT Required
Degradation of soluble cBOD	0.3
Floc formation (domestic wastewater)	1
Solubilization of particulte and collodial cBOD	2
Biological phosphorus removal	2
Floc formation (industrial wastewater)	3
Degradation of xenobiotic cBOD	5
Maturation of floc particles	10
Nitrification (temperature-dependent)	3–15

TABLE 8.2 SRT (Approximate Days) Required for the Initiation of Significant Biological Activities in the Anaerobic Digester

Biological Activity	Approximate SRT Required
Fermentation (acidogenesis), acids and alcohol production	0.5
Solubilization of particulate and colloidal cBOD	0.5
Methane production (H_2 utilizing methanogens)	0.5
Methane production (acetate utilizing methanogens)	3
Fermentation of volatile fatty acids	4
Fermentation of long-chain fatty acids	4
Solubiization of lipids	6

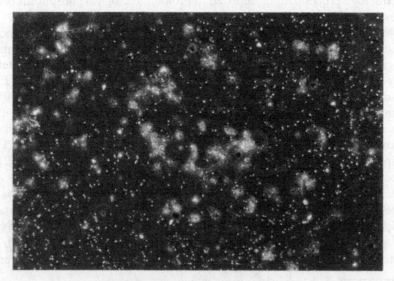

FIGURE 8.1 Dispersed bacterial growth.

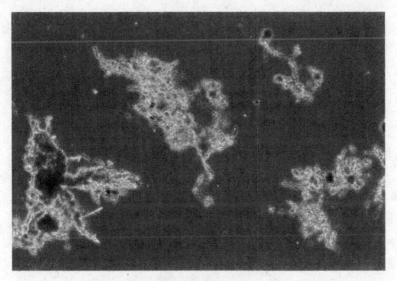

FIGURE 8.2 *Flocculated bacteria or floc particle.*

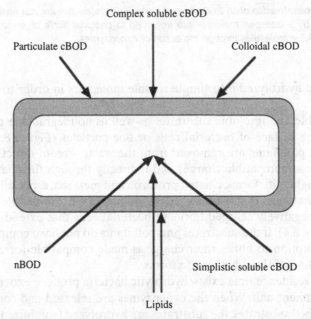

FIGURE 8.3 *Absorption and adsorption of substrate. Only simplistic soluble substrates, cBOD, nBOD (ionized ammonia and nitrite), and some lipids, are easily absorbed by bacterial cells. Other substrates are adsorbed to bacteria and are solublized through the action of exoenzymes or conveyed into the cell with specific transport molecules (proteins).*

an immediate demand for nutrients and an electron carrier molecule such as free molecular oxygen or nitrate (NO_3^-). Adsorbed substrates are those that are insoluble or poorly soluble, are complex in structure, and do not enter bacterial cells. Examples of these adsorbed substrates include starches such as cellulose, lipids (fats and oils), proteins, and even disaccharides such as lactose and maltose. These sub-

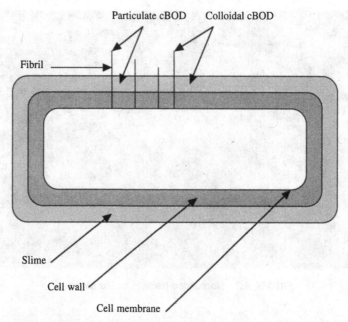

Particulate cBOD   Colloidal cBOD

Fibril

Slime

Cell wall

Cell membrane

FIGURE 8.4 *Substrate adsorption to fibrils and slime. Substrates that are not absorbed or carried into a bacterium by a transport molecule are adsorbed to bacterial fibrils or slime where they are degraded into soluble molecules through the action of exoenzymes.*

strates must be hydrolyzed into simple soluble molecules in order to enter bacterial cells.

Poorly soluble and insoluble substrates as well as nondegradable pollutants are adsorbed to the surface of bacterial cells or floc particles (Figure 8.3). Adsorbed substrates and pollutants are removed from the waste stream directly by electrochemical process (compatible charge) and indirectly through the coating action of secretions of higher life forms, ciliated protozoa and metazoa, especially rotifers and free-living nematodes. If substrates and pollutants have compatible charge, they attach to the negatively charged fibrils of bacterial cells that extend into the bulk solution (Figure 8.4). If the substrates and pollutants do not have compatible charge for direct adsorption to fibrils, their charge is made compatible for adsorption by the coating action of these higher life forms.

If sufficient residence time exists, hydrolytic bacteria produce exoenzymes in the biological treatment unit. When the exoenzymes are released and come in contact with the adsorbed substrates, the substrates are hydrolyzed (solublized) into smaller and soluble substrates that then are absorbed by they hydrolytic bacteria and non-hydrolytic bacteria in the biological treatment unit (Figure 8.5). Once absorbed, the small and soluble substrates are degraded by endoenzymes.

Hydrolytic bacteria consist of a consortia of Gram-positive, rod-shaped, facultative anaerobic bacteria and anaerobic bacteria that break down poorly soluble and insoluble complex carbohydrates, lipids, and proteins into simple and soluble sugars, fatty acids and glycerine ($CH_2OHCHOHCH_2OH$), and amino acids, respectively. These soluble substrates are available for absorption and degradation by numerous bacteria.

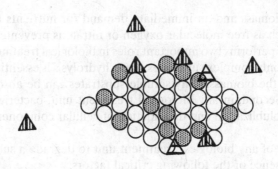

○ Non-hydrolytic bacteria

◉ Hydrolytic bacteria

▲ Particulate cBOD

△ Particulate cBOD undergoing hydrolysis

FIGURE 8.5 *Hydrolysis and absorption of substrate. Particulate and colloidal cBOD that is adsorbed to the surface of bacterial cells is hydrolyzed (solublized) by the production and release of exoenzymes from hydrolytic (exoenzyme-producing) bacteria. Once particulate and colloidal cBOD have been hydrolyzed, the soluble substrates produced through hydrolysis are absorbed and degraded by not only the hydrolytic bacteria but also the nonhydrolytic bacteria. The hydrolysis and absorption of substrate occurs in a limited amount in the activated sludge process (flocculated bacteria) and in a large amount in the anaerobic digester due to the higher solids retention time and larger diversity of hydrolytic bacteria in the anaerobic digester.*

TABLE 8.3   Examples of Exoenzymes

Exoenzyme (Specificity or Compounds)	Function
Amylase (carbohydrate)	Converts starch to maltose
Caseinase (milk protein)	Converts milk protein to peptides and amino acids
Cellulase (carbohydrate)	Converts cellulose to cellobiose
Gelatinase (gelatin)	Converts gelatin to peptides and amino acids
Lactase (disaccharide/sugar)	Converts lactose to glucose and galactose
Lipase (lipids)	Converts lipids (fat) to glycerol and fatty acids
Maltase (disaccharide/sugar)	Converts maltose to two glucose molecules
Proteinases (protein)	Converts proteins to peptides and amino acids
Sucrase (disaccharide/sugar)	Converts sucrose to glucose and fructose

Hydrolysis is the addition of water ("hydro") to complex molecules by bacteria to split ("lysis") unique chemical bonds in the complex molecules, thereby permitting the production and release of simple and soluble molecules. The addition of water and breakage of chemical bonds is catalyzed by exoenzymes such as cellulase (hydrolyze starches or carbohydrates), lipases (hydrolyze lipids), and proteases (hydrolyze proteins) (Table 8.3). Hydrolysis occurs slowly. The more slowly a complex molecule is hydrolyzed, the more slowly soluble substrates are made available to the biomass and an immediate demand for nutrients and electron carrier molecules such as free molecular oxygen or nitrate is prevented.

Hydrolysis performs two important roles in biological treatment units. First, many units receive only complex substrates. Here, hydrolysis is essential to provide soluble substrates to the biomass. Only soluble substrates can be absorbed and degraded by bacteria. Second, in any biological treatment unit, bacteria die and hydrolysis permits the solublization and degradation of cellular components rather than their accumulation.

The ability of any biological treatment unit to degrade a substrate is dependent upon the presence of the following critical factors:

- A diverse population of bacteria and enzymes
- An adequate number of enzymes
- Acceptable operational conditions including MCRT
- The molecular structure of the substrate

Bacteria degrade first the highly soluble and simple substrates, then the insoluble and complex substrates. A substrate such as sugar is readily degradable because the following conditions occur in the biological treatment unit:

- All the necessary enzymes for the degradation of the substrate are initially present in adequate numbers.
- The rate of all biochemical reactions that are necessary for complete degradation of the substrate are optimum.
- An adequate residence time is available for the degradation of the substrate.

A substrate such as chitin or wax is poorly degradable because the following conditions occur in the biological treatment unit:

- All the necessary enzymes for the degradation of the substrate are not initially present in adequate numbers.
- The rate of all biochemical reactions that are necessary for complete degradation of the substrate are not optimum.
- An adequate residence time is not available for the degradation of the substrate.

Poorly degradable substrates have chemical structures for which no organism can produce an enzyme for its degradation. Poorly degradable substrates also are known as incompletely degradable substrates or nonbiodegradable substrates.

9

Bacterial Growth

Biological, wastewater treatment plants are simply biological amplifiers. The plants permit organisms (biomass or sludge), primarily bacteria, to increase in number by using the pollutants (substrates and nutrients) in the wastewater and converting them to new organisms (biomass or sludge) and nonpolluting wastes and less polluting wastes (Table 9.1). Nonpolluting wastes do not contribute to operational or environmental problems. Less polluting wastes are not as harmful as the original pollutants but may contribute to operational and/or environmental problems.

The carbon and energy sources (wastes) in the wastewater for bacterial growth are referred to as substrates. Therefore, the degradation of substrates for bacterial growth is referred to as substrate utilization. The increase in the quantity of bacteria (sludge) per unit of substrate utilized is known as the growth yield or sludge yield (Table 9.2).

Bacterial cells do grow; but as soon as a cell or mother cell has approximately doubled in size, it divides into two offspring or daughter cells. This form of growth or reproduction is known as binary fission (Figure 9.1. In some bacteria, incomplete separation of the daughter cells produces different arrangements or patterns of cellular growth such as filaments and tetrads.

Because bacterial cells grow in size only to divide into two cells, microbial growth in the microbiology laboratory can be defined in terms of the number of cells produced through division. However, in the wastewater treatment plant the number of cells produced in a biological treatment unit such as the activated sludge process cannot be determined due to the following difficulties:

- There is no growth medium that would permit the growth and enumeration of all bacteria in any biological treatment unit.

Wastewater Bacteria, by Michael H. Gerardi
Copyright © 2006 John Wiley & Sons, Inc.

TABLE 9.1 Examples of Nonpolluting Wastes and Less Polluting Wastes Produced from Aerobic[a] and Anoxic[b] Degradation of Proteins[c]

Form of Degradation	Wastes		
	Nonpolluting	Less Polluting	Concern with Less Polluting Waste
Aerobic	CO_2, H_2O	$H_2PO_4^-$	Algal blooms in receiving waters
		NH_4^+	Nitrification and oxygen demand
		SO_4^{2-}	Reduction and H_2S production
Anoxic	CO_2, H_2O, N_2, N_2O	$H_2PO_4^-$	Algal blooms in receiving waters
		HS^-	Growth of filamentous organisms
		NH_4^+	Nitrification and oxygen demand

[a] Aerboic degradation uses free molecular oxygen.
[b] Anoxic degradation uses nitrate (NO_3^-).
[c] Proteins contain C, H, O, N, S, and P.

TABLE 9.2 Examples of Sludge Yields (Pounds of Sludge per Pound of cBOD Removed)

Quantity of cBOD (lb)	Type of cBOD	Degradation	Approximate Sludge Yield (lb)
1	Sugar	Aerobic	0.6
1	Sugar	Anoxic	0.4
1	Proteins	Aerobic	0.4

- Due to the large diversity of bacteria in a biological treatment unit, there are many different generation (reproduction) times. Some generation times may be as short as 15 minutes (organotrophs) and as long as 15 days (nitrifying bacteria).
- Erratic wasting rates produce "pockets" of young growth and old growth with different generation times.
- It is time-consuming, labor intense, and expensive to use a variety of techniques to determine the number of different bacteria in a biological treatment unit.

Due to the difficulties encountered in enumerating bacteria in wastewater treatment plants, volatile solids concentrations are used to estimate the size or mass of bacterial populations. Because bacteria are organic in composition, they are volatilized in a muffle furnace at 550°C. Therefore, an increase in volatile solids concentration is considered to be an increase in the mass of the bacterial population, while a decrease in volatile solids concentration is considered to be a decrease in the mass of the bacterial population. The mixed liquor volatile suspended solids (MLVSS) concentration is used in the activated sludge process to estimate the mass of the bacterial population.

The use of volatile solids such as the MLVSS to determine the mass of the bacterial population provides only an estimate of the mass. There are several operational concerns that impact the composition of the volatile solids that influence an estimate of the bacterial population. These concerns are as follows:

- Volatile solids concentration can change significantly due to the accumulation or lost of compounds such as fats, oils, and grease, insoluble polysaccharides

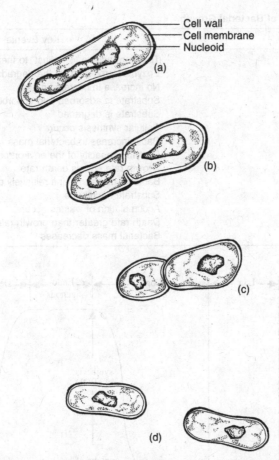

Cell wall
Cell membrane
Nucleoid

(a)

(b)

(c)

(d)

FIGURE 9.1 *Binary fission in a bacterium. During binary fission the nucleoid elongates (a) and then divides (b). Division of the nucleoid is accompanied by the formation of transverse septum by the cell wall (b). After the transverse septum is complete (c), two new cells or daughter cells are produced (d).*

produced through slug discharges of soluble cBOD, and stored starches produced during a nutrient deficiency.

- Protozoa and metazoa may contribute to up to 5% of the volatile content of the solids.
- Volatile solids concentration does not indicate if the bacterial population has experienced inhibition or toxicity.

Bacterial cells in biological treatment units reproduce mainly by binary fission; that is, each mother cell produces two daughter cells. The growth of the bacterial population is defined as an increase in the mass of bacteria, and the growth of the bacteria can occur as a batch culture (closed system) or a continuous culture (open system). In a batch culture there is a limited quantity of substrate. An example of a batch culture occurs in an activated sludge process when an aeration tank is taken "off-line" but is not drained and continues to be aerated and mixed. An example of a continuous culture in an activated sludge process occurs when an aeration tank remains "on-line" and continuously receives substrates.

TABLE 9.3 Phases of Bacterial Growth

Phase	Key Events
Lag	Time for bacteria to "adjust" to their new environment
	Enzymes produced for the degradation of substrate
	No increase in bacterial mass
Log	Substrate is adsorbed and absorbed
	Substrate is degraded
	Cellular synthesis occurs
	Rapid increase in bacterial mass
Endogenous	Carrying capacity of the environment is reached
	Growth rate equals death rate
	Bacterial mass remains relatively constant
Death	Substrate decreases
	Accumulation of wastes occurs
	Death rate greater than growth rate
	Bacterial mass decreases

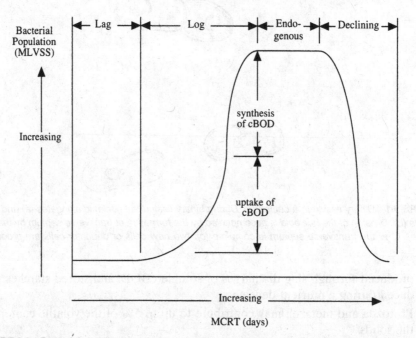

FIGURE 9.2 *Bacterial growth curve, batch culture or batch reactor. In a batch culture, substrates are provided only once [e.g., an aeration tank taken off-line or a sequential batch reactor (SBR)].*

There are four distinct phases of bacterial growth in batch cultures and continuous cultures (Table 9.3). These phases are lag, log, endogenous, and death or decline. Only the death phase in the growth curves for batch cultures (Figure 9.2) and continuous cultures (Figure 9.3) differs.

LAG PHASE OF GROWTH

During lag phase of growth, bacteria are active but do not reproduce. The bacteria are synthesizing enzymes to degrade substrates in their new environment. The

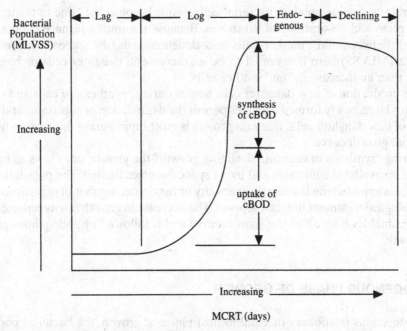

FIGURE 9.3 *Bacterial growth curve, continuous culture or continuous reactor. In a continuous culture, substrates are provided continuously (e.g., an on-line aeration tank).*

degradation of substrates provides the carbon and energy needed for reproduction. Lag phase of growth occurs in biological treatment units during start-up and recovery from toxicity. Sudden shock loadings with substrates not previously contained in the wastewater can produce a short lag phase of growth for a few hours. This is due to the fact that bacteria do not have the appropriate enzymes synthesized to degrade the new substrates in the shock loading condition. If the bacteria have the ability to synthesize the enzymes when expose to the new substrates, the enzymes are produced (substrate-induced enzyme synthesis).

In addition to the synthesis of enzymes, bacteria are adjusting to their new environment in the biological treatment unit. For example, the biological treatment unit possesses different biological, chemical, and physical conditions than the environment (soil, water, or gastrointestinal tract of humans) where the bacteria previously had inhabited. The length of lag phase of growth is determined by (1) the conditions of the new environment and (2) the species of bacteria present. Bacteria in a minimal nutrient environment take longer to adjust than bacteria in a rich nutrient environment. Some bacteria adjust quickly in one hour, while other bacteria take several days to adjust.

LOG PHASE OF GROWTH

Log phase of growth also is known as exponential or logarithmic phase, because the bacterial population grows at a logarithmic rate. There are three significant portions to the log phase of growth. These portions are (1) substrate uptake, (2) synthesis of cells and rapid growth, and (3) synthesis of cells and declining growth.

During "substrate uptake," bacterial cells simple become "fat" due to the uptake (adsorption and absorption) of substrates. Because the muffle furnace cannot distinguish between "fat" mother cells and daughter cells, the increase in volatile content (MLVSS) here is assumed to be an increase in daughter cells or biomass rather than an increase in "fat" mother cells.

The production of new daughter cells begins during "synthesis of cells and rapid growth." Here, newly formed enzymes permit the degradation of substrates and synthesis of new daughter cells. Bacteria growth is most rapid during this section of the bacterial growth curve.

During "synthesis of cells and declining growth" the growth rate slows as bacteria use up available substrates and living space becomes limited. The population of bacteria is approaching the carrying capacity or maximum number of organisms that the biological treatment unit can support. The decrease in growth rate is represented by a gradual leveling off of the growth curve and is followed by endogenous phase of growth.

ENDOGENOUS PHASE OF GROWTH

At endogenous (stationary or equilibrium) phase of growth, the bacterial population has reached the carrying capacity of the biological treatment unit. The bacteria cannot grow indefinitely due to (1) the lack of an ever-increasing quantity of substrates, (2) the lack of electron acceptors such as free molecular oxygen or nitrate, and (3) production and accumulation of toxic metabolic wastes. There is no net increase in bacterial growth. Cellular growth is balanced by cellular death during endogenous phase of growth.

DEATH OR DECLINE PHASE OF GROWTH

During the death or decline phase of growth, the death rate for bacteria exceeds the growth rate for bacteria. In batch cultures (Figure 9.2) the negative slope of the death phase or decline phase of growth is more dramatic or steep than the negative slope in continuous cultures (Figure 9.3). Batch cultures are limited with respect to the quantity of substrates received, while continuous cultures receive some substrates at all times. Therefore, the death rate of bacteria in the continuous cultures is not as dramatic as the death rate in the batch cultures. However, the quantity of substrates received by continuous cultures varies throughout the week and time of day. Therefore, the negative slope of death or decline phase of growth varies according to substrate loading conditions.

ANABOLISM VERSUS CATABOLISM

Substrate that is absorbed by bacterial cells is degraded to provide carbon and energy for cellular growth and cellular activity. When substrate is used for cellular synthesis, small molecules are joined together to form large molecules and cellular growth occurs. This is referred to anabolism, and sludge production increases. When

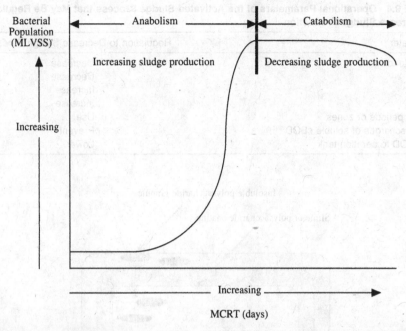

FIGURE 9.4 *Bacterial growth curve, occurrence of anabolism and catabolism. Anabolic reactions favor increased sludge production, while catabolic reactions favor decreased sludge production. Operational conditions that promote anabolic reactions include (1) excess loss of solids from the secondary clarifiers, (2) overwasting or erratic wasting rates, (3) recovery from toxicity, (4) recovery from washout (excess I/I), and (5) slug discharge of soluble cBOD. Operational conditions that promote catabolic reactions include (1) maintenance of a low F/M and high MCRT, (2) decrease in cBOD loading, (3) rotating aeration tanks "on-line" and "off-line," and (4) increasing the HRT of the aeration tanks.*

substrate is used for energy production, large molecules are degraded to smaller molecules and cellular energy is obtained. This is referred to as catabolism, and sludge production decreases.

On the bacterial growth curve (Figure 9.4), anabolism or increase in sludge production occurs before endogenous phase of growth. Catabolism occurs with the onset of endogenous phase of growth. Anabolism always occurs after the start-up of a wastewater treatment plant.

Because sludge thickening, dewatering, and disposal costs are a major operational expense for most wastewater treatment plants, these plants should be operated to favor catabolic conditions or a decrease in sludge production. Anabolism should occur once, immediately after start-up. However, there are operational conditions that do favor anabolism and an increase in sludge production. These operational or anabolic conditions including the following:

- Excess loss of solids from the secondary clarifiers in the activated sludge process
- Overwasting of solids or erratic wasting rates
- Recovery from inhibition or toxicity
- Recovery from washout
- Slug discharge of soluble cBOD

TABLE 9.4 Operational Parameters of the Activated Sludge Process that May Be Regulated to Decrease Sludge Production

Parameter	Regulation to Decrease Sludge Production
cBOD loading	Decrease
F/M	Decrease
HRT	Increase
MCRT	Increase
Anoxic periods or zones	Use
Slug discharges of soluble cBOD	Prevent
TSS:BOD to aeration tank	Lower

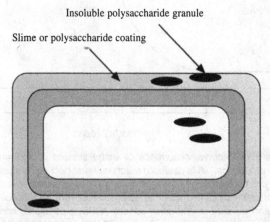

FIGURE 9.5 *Stored food. Large quantities of food often are stored by bacteria as starch granules or slime (polysaccharide coating). Starch granules may be located within the bacterial cell or outside the bacterial cell in the slime.*

During endogenous phase of growth, the activated sludge process can be monitored and regulated to decrease sludge production (Table 9.4). By decreasing the substrate (cBOD) loading to the wastewater treatment plant, less bacterial growth (sludge production) occurs and food stored by bacteria is solubilized and degraded (Figure 9.5). A significant decrease in cBOD loading may be obtained by reducing the quantity of fats, oils, and grease discharged to the sewer system. By using appropriate-sized grease traps on cafeteria, commercial, and industrial waste streams and capturing and disposing of fats, oils, and grease at approved landfills, decreased cBOD loading can be obtained. In addition to decreased cBOD loading, the following benefits are obtained for the sewer system and the wastewater treatment plant:

- Decreased occurrences of saponification of fats, oils, and grease in the sewer system, resulting in less interruption of sewer service and decreased maintenance of the sewer system
- Decrease in volatile content of the mixed liquor suspended solids (MLSS), resulting in improved settleability
- Decrease in the growth of foam-producing filamentous organisms (*Microthrix parvicella* and Nocardioforms) that use fats, oils, and grease as a substrate

By lowering the F/M of the activated sludge process, less substrate is available to all bacteria. Therefore, decreased cellular growth results, bacteria consume stored food, and bacteria die. When lowering the F/M, caution should be used to prevent the undesired growth of low F/M filamentous organisms such as *Haliscomenobacter hydrossis*, *Microthrix parvicella*, Nocardioforms, type 021N, and type 0041.

By increasing the hydraulic retention time (HRT) within the aeration tanks, solubilization of some particulate and colloidal cBOD may occur. Once solubilized, these substrates can be transformed to bacterial cells that represent less weight than the original particulate and colloidal cBOD. However, by increasing the HRT and solubilizing substrates, an increase in dissolved oxygen demand may occur, with the occurrence of nitrification.

By increasing the MCRT a large number of "older" bacteria are maintained in the activated sludge process. These bacteria favor catabolic events (decreased sludge production) over anabolic events (increased anabolic events). However, caution should be exercised when increasing the MCRT. An increase in MCRT may contribute to the following operational concerns:

- Significant endogenous respiration that consumes large quantities of dissolved oxygen
- Undesired growth of high MCRT filamentous organisms such as *Microthrix parvicella*, type 0041, type 0092, and type 1851
- Undesired growth of nitrifying bacteria and nitrification

The use of anoxic periods (use of nitrate) to degrade cBOD results in decreased sludge production. For example, based upon sludge yields, one pound of sugar degraded with free molecular oxygen results in the production of approximately 0.6 pound of sludge (bacterial cells), while one pound of sugar degraded with nitrate results in the production of approximately 0.4 pound of sludge (bacterial cells). When using anoxic periods, the presence of nitrate ions should never be exhausted. If nitrate is exhausted, septicity occurs.

Slug discharges of soluble cBOD should be prevented. These discharges usually are easily degradable and result in the rapid proliferation of bacterial cells. This rapid and young growth is accompanied with the production of a copious quantity of buoyant and insoluble polysaccharides. Discharges of highly soluble cBOD should be identified and prevented.

By lowering the TSS (total suspended solids):BOD ratio to the aeration tank influent, decreased sludge production occurs. Less suspended solids to the aeration tank results in less solids (sludge) from the aeration tank. By improving primary clarifier efficiency, a reduction in the quantity of suspended solids discharged to the aeration tank can be reduced. For example, additional clarifiers can be placed "online," clarifiers can be pumped more frequently, and a polymer can be added to the primary clarifier influent to remove more solids.

Part III

Nitrogen, Phosphorus, and Sulfur Bacteria

10

Nitrifying Bacteria

Nitrogen (N) is an essential nutrient of all living organisms. In the amino group (—NH$_2$) nitrogen is in the –3 oxidation state or valence and is incorporated into amino acids such as glycine (CH$_2$NH$_2$COQH) (Figure 10.1). Amino acids are used by organisms to build proteins. Proteins then are used to build structural materials, enzymes, and genetic materials. Amino acids, proteins, and compounds built with proteins are organic nitrogen compounds. These compounds are found in fecal waste and food waste discharged to municipal wastewater treatment plants.

While most amino acids are simple in structure and soluble in water, proteins are complex in structure and insoluble in water. Proteins are colloids and possess a relatively large surface area and remain suspended in water.

Within the sewer system, simple amino acids undergo deamination through bacterial activity (Figure 10.2). Deamination results in the release of amino groups and the production of reduced nitrogen. There are two forms of reduced nitrogen: ammonia (NH$_3$) and ammonium ion or ionized ammonia (NH$_4^+$). The quantity of each form of reduced nitrogen that is produced is pH-dependent (Figure 10.3).

Ammonia is toxic and is released to the atmosphere from the sewer system or biological treatment unit through turbulence (aeration or mixing action). Ionized ammonia is nontoxic and is used by bacteria as their preferred nitrogen nutrient. At pH values lower than 9, most of the reduced nitrogen is present as ionized ammonia.

Another significant organic nitrogen compound is urea (H$_2$NCONH$_2$). Urea is a major component of urine that is discharged to municipal wastewater treatment plants. Urea undergoes hydrolysis within the sewer system through bacterial activity. Hydrolysis is the addition of water to a compound resulting in the "splitting of the compound into at least two simpler compounds. Hydrolysis of urea results in the production of reduced nitrogen (Equation 10.1).

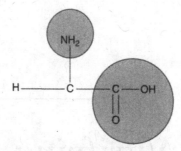

FIGURE 10.1 *Glycine (aminoethanoic acid). Glycine is the simplistic amino acid. Like all amino acids, glycine possesses a carboxyl group (—COOH) and an amino group (—NH₂). The carboxyl group is located on a terminal carbon unit, while the amino group is located on the carbon unit immediately adjacent to the carboxyl group.*

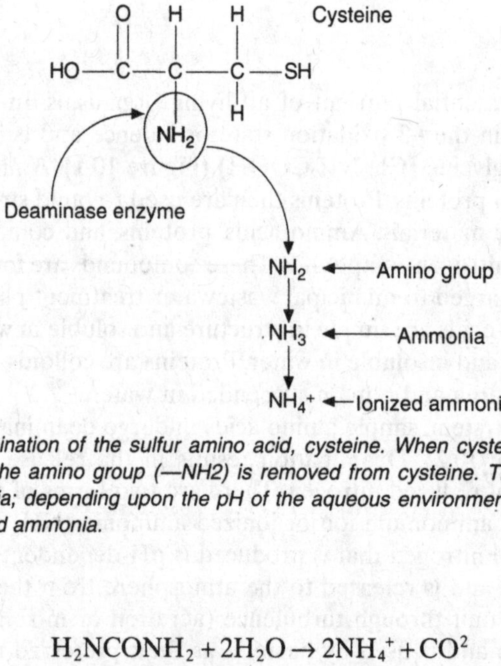

FIGURE 10.2 *Deamination of the sulfur amino acid, cysteine. When cysteine is "attacked" by the deaminase enzyme, the amino group (—NH2) is removed from cysteine. The liberated amino acid quickly forms ammonia; depending upon the pH of the aqueous environment, the ammonia may be ionized to form ionized ammonia.*

$$H_2NCONH_2 + 2H_2O \rightarrow 2NH_4^+ + CO^2 \tag{10.1}$$

The influent to municipal wastewater treatment plants contains organic nitrogen compounds and inorganic nitrogen in the form of ionized ammonia. Approximately 60% of the nitrogenous wastes to the treatment plants is in the organic forms, and approximately 40% of the nitrogenous wastes is in the inorganic form. Most municipal wastewater treatment plants have an influent, ionized ammonia concentration of 25–30 mg/liter.

Additional nitrogenous wastes that may enter municipal wastewater treatment plants include polymers that are used at wastewater treatment plants and industrial wastewaters (Table 10.1). Some of these industrial wastewaters contain not only ionized ammonia and organic nitrogen compounds but also nitrite (NO_2^-) and nitrate (NO_3^-).

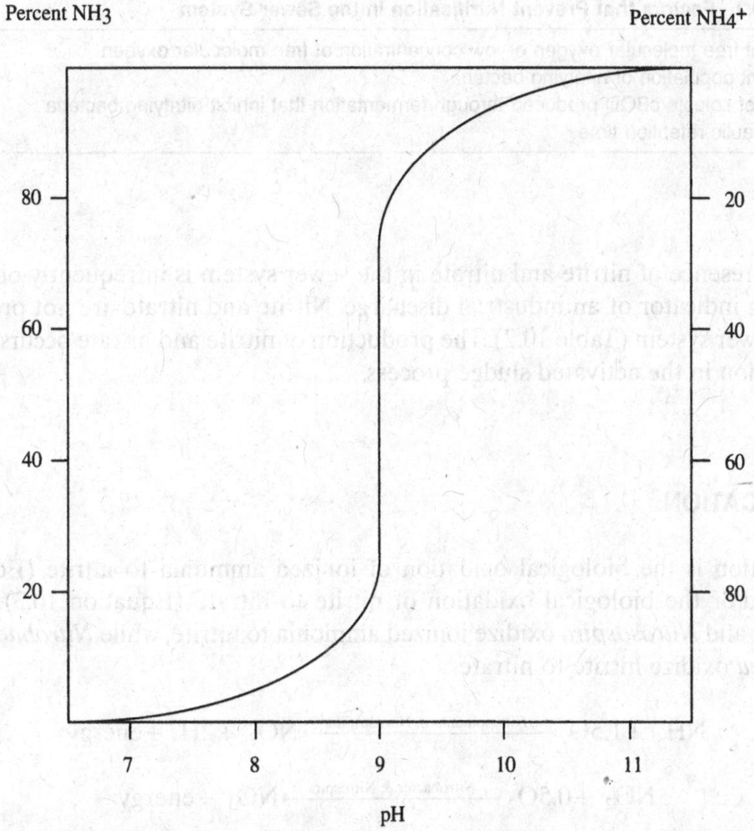

Percent NH3   Percent NH4+

FIGURE 10.3 *Distribution of ammonia and ionized ammonia in the mixed liquor.*

TABLE 10.1 Industrial Discharges of Ionized Ammonia, Nitrite Ions, and Nitrate Ions

	Nitrogenous Compound		
Industrial Discharge	NH_4^+	NO_2^-	NO_3^-
Automotive	X		
Chemical	X		
Coal	X		
Corrosion inhibitor		X	
Fertilizer	X		
Leachate	X		
Leachate (pretreated)		X	X
Livestock	X		
Meat	X		
Meat (flavoring)			X
Meat (preservative)		X	
Ordnance	X		
Petrochemical	X		
Pharmaceutical			X
Primary metal	X		
Refineries	X		
Steel	X	X	X
Tanneries	X		

TABLE 10.2   Factors that Prevent Nitrification in the Sewer System

Absence of free molecular oxygen or low concentration of free molecular oxygen
Insignificant population of nitrifying bacteria
Presence of soluble cBOD produced through fermentation that inhibit nitrifying bacteria
Short hydraulic retention time

The presence of nitrite and nitrate in the sewer system is infrequently observed and is an indicator of an industrial discharge. Nitrite and nitrate are not produced in the sewer system (Table 10.2). The production of nitrite and nitrate occurs during nitrification in the activated sludge process.

NITRIFICATION

Nitrification is the biological oxidation of ionized ammonia to nitrite (Equation 10.2) and/or the biological oxidation of nitrite to nitrate (Equation 10.3). *Nitrosomonas* and *Nitrosospira* oxidize ionized ammonia to nitrite, while *Nitrobacter* and *Nitrospira* oxidize nitrite to nitrate.

$$NH_4^+ + 1.5O_2 \xrightarrow{\text{Nitrosomonas \& Nitrosospira}} NO_2^- + 2H^+ + \text{energy} \qquad (10.2)$$

$$NO_2^- + 0.5O_2 \xrightarrow{\text{Nitrobacter \& Nitrospira}} NO_3^- + \text{energy} \qquad (10.3)$$

Nitrifying bacteria oxidize ionized ammonia and nitrite in order to obtain energy for cellular activity including reproduction. In the activated sludge process, this reproduction results in an increase in sludge. Carbon needed by nitrifying bacteria is obtained from bicarbonate alkalinity (HCO_3^-). Because nitrifying bacteria obtain very little energy from the oxidation of ionized ammonia and nitrite, bacterial growth or sludge production is relatively small. Approximately 0.06 pound of nitrifying bacteria or sludge is produced for every pound of ionized ammonia oxidized to nitrate.

Nitrifying bacteria are chemolithoautotrophs. As chemolithoautotrophic bacteria, they obtain their cellular energy by oxidizing a mineral such as nitrogen, and they obtain their carbon for cellular synthesis by consuming inorganic carbon. Inorganic carbon does not contain hydrogen (H). The inorganic carbon source for autotrophic nitrifying bacteria is carbon dioxide (CO_2). Carbon dioxide is consumed in the form of bicarbonate alkalinity (HCO_3^-)—that is, carbon dioxide dissolved in water to form carbonic acid (H_2CO_3) that dissociates to form bicarbonate alkalinity (Equation 10.4).

$$H_2CO_3 \leftrightarrow H^+ + HCO_3^- \qquad (10.4)$$

Nitrifying bacteria are very efficient in oxidizing ionized ammonia and nitrite because they possess (1) unique nitrifying enzyme systems and (2) cytomembranes (Figure 10.4). Cytomembranes are an in-folding of the cellular membrane that

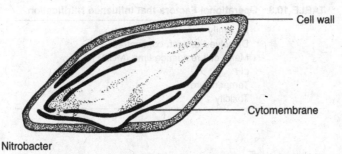

Nitrobacter

FIGURE 10.4 *Cytomembranes in nitrifying bacteria.*

provide an increase in the surface area of the cell membrane upon which nitrification can occur.

Nitrifying bacteria are free-living organisms and are found in the soil and water. They enter wastewater treatment plants through inflow and infiltration. Because nitrifying bacteria are strict aerobes, they live in the top 1–2 inches of soil.

Due to the relatively small quantity of energy obtained from the oxidation of ionized ammonia and nitrite, nitrifying bacteria reproduce very slowly. Under optimal conditions, the generation time for nitrifying bacteria is approximately 8–10 hours. Under the harsh conditions of an activated sludge process, the generation time for nitrifying bacteria is approximately 2–3 days. Therefore, in activated sludge processes, relatively high mean cell residence times (MCRT) are required to establish a population of nitrifying bacteria that are capable of effective nitrification.

The activity and generation time of nitrifying bacteria are temperature-dependent. With increasing temperature nitrifying bacteria become more active and reproduce more quickly. Nitrifying bacteria are active and reproduce over a range of temperature values from 5°C to 40°C. However, the maximum temperature for nitrification in the activated sludge process is considered to be 30°C. This temperature produces the maximum activity and shortest generation time for *Nitrosomonas*, one of the major nitrifying bacteria.

Nitrifying bacteria are poor floc-forming organisms. Their incorporation into ftoc particles is achieved largely through compatible charges between the nitrifying bacteria and the floc particle and their adsorption to the floc particles through coating action of higher life forms. Ciliated protozoa, rotifers, and free-living nematodes, several of the higher life forms in the activated sludge process, release secretions that coat the surface of nitrifying bacteria and renders the surface compatible for adsorption to the floc particles.

Due to the long generation time of nitrifying bacteria and their small population growth (sludge yield) from the oxidation of energy substrates (NH_4^+ and NO_2^-), nitrifying bacteria usually represent <10% of the bacterial population in the activated sludge process. Because nitrifying bacteria are strict aerobes, they are found mostly on the perimeter of the floc particles.

Although the activity and maximum population size of nitrifying bacteria are dependent upon the quantity of energy substrates that are available, there are several operational factors that influence the activity and population size of nitrifying bacteria and the ability of the activated sludge process to successfully nitrify

TABLE 10.3   Operational Factors that Influence Nitrification

Alkalinity
Dissolved oxygen concentration
Mean cell residence time (MCRT)
pH
Temperature
Toxicity

TABLE 10.4   Temperature and MCRT Recommended for Nitrification

Temperature (°C)	MCRT (days)
30	7
25	10
20	15
15	20
10	30

TABLE 10.5   Temperature and Nitrification

Temperature (°C)	Effect on Nitrification
30	Optimum temperature for nitrification
15	Approximately 50% of optimum nitrification
10	Approximately 20% of optimum nitrification
5	Nitrification ceases

(Table 10.3). These factors include alkalinity and pH, dissolved oxygen concentration, MCRT and temperature, and toxicity. The most important factors are MCRT and temperature.

The most critical factors affecting the activity and population size of hitrifying bacteria and the success of nitrification in the activated sludge process are MCRT and temperature. Because there is an indirect relationship between temperature and activity of nitrifying bacteria, increasing MCRT is required with decreasing temperature (Table 10.4). With decreasing wastewater temperature, nitrifying bacteria become less active, and nitrification efficiency decreases. To compensate for the lost of activity and to improve nitrification efficiency, the number of nitrifying bacteria must be increased. An increase in the number of nitrifying bacteria requires an increase in the mixed liquor volatile suspended solids (MLVSS) and MCRT.

The most critical temperature value with respect to process control of nitrification is 15°C (Table 10.5). With decreasing wastewater temperature, approximately 50% of the ability of the activated sludge process to nitrify is lost at 15°C, unless appropriate operational measures are implemented to maintain effective nitrification (Table 10.6).

Adequate alkalinity is essential for successful nitrification. At least 50 mg/liter of alkalinity should be present in the mixed liquor effluent after complete nitrification to ensure the presence of adequate alkalinity. After complete nitrification, the mixed

TABLE 10.6 Operational Measures Available for Maintaining Effective Nitrification During Cold Wastewater Temperature

Operational Measure	Effect
Addition of bioaugmentation products	Bacterial cultures rapidly remove cBOD and improve or initiate nitrification
Increase aeration tank dissolved oxygen concentration	Promotes rapid cBOD removal and improved nitrification
Increase hydraulic retention time (HRT)	Provides more time for nitrification
Increase primary clarifier efficiency	Removes more colloidal and particulate cBOD and lowers dissolved oxygen demand in the aeration tank
Install biological holdfast (ringlace) system	Increases bacterial populations for removing cBOD quickly and nitrifying without overloading the secondary clarifier

liquor effluent contains <1 mg/liter ionized ammonia (NH_4^+) and <1 mg/liter nitrite (NO_2^-).

There are two biochemical reactions that are responsible for the loss of alkalinity during nitrification. The minor reaction is the used of bicarbonate alkalinity as the carbon substrate for the synthesis of cellular materials and reproduction. The major reaction is the production of free nitrous acid (HNO_2) that destroys alkalinity.

When ionized ammonia is oxidized to nitrite during nitrification, hydrogen protons (H^+) are produced (Equation 10.5). When hydrogen protons combine with nitrite, free nitrous acid is produced (Equation 10.6). As nitrous acid is produced, alkalinity is destroyed, and the pH of the mixed liquor drops.

$$NH_4^+ + 1.5O_2 \rightarrow 2H^+ + NO_2^- + H_2O \qquad (10.5)$$

$$H^+ + NO_2^- \rightarrow HNO_2 \qquad (10.6)$$

Although alkalinity is lost during nitrification, some alkalinity is returned naturally to the activated sludge process through deamination of organic nitrogen compounds and denitrification (Equation 10.7). Deamination of organic nitrogen compounds results in the production of ionized ammonia. Ionized ammonia represents an increase in alkalinity. Through denitrification (i.e., the use of nitrate to degrade soluble cBOD), alkalinity is produced in two forms. First, the production of hydroxyl ions (OH^-) returns alkalinity directly to the process. Second, the release of carbon dioxide that dissolves in the wastewater returns alkalinity indirectly through the formation of the bicarbonate ion (HCO_3^-).

$$6NO_3^- + 5CH_3OH \rightarrow 3N_2 + 5CO_2 + 7H_2O + 6OH^- \qquad (10.7)$$

Nitrifying bacteria are active over a wide range of pH values, 5 to 8.5 (Table 10.7). Although the optimum pH range for nitrification is 7.3 to 8.5, most activated sludge processes nitrify at a near neutral pH value, 6.8 to 7.2. At pH values greater than 7.3, undesired operational conditions may occur in the activated sludge process. These undesired operational conditions include the following:

TABLE 10.7 pH and Nitrification

pH	Effect on Nitrification
4.0–4.9	Nitrifying bacteria present but inactive; limited nitrification occurs through the activity of organotrophic bacteria
5.0–6.7	Nitrifying bacteria are active but activity is sluggish
6.8–7.2	Desired pH range for nitrification in the activated sludge process
7.3–8.0	Rate of nitrification assumed be constant
8.1–8.5	Optimum pH range for nitrification (e.g., in laboratory work with nitrifying bacteria only)

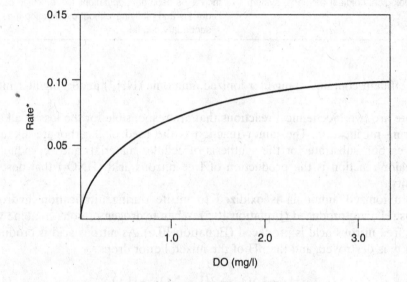

*Pound of ionized ammonia oxidized per pound ML VSS per day

FIGURE 10.5 *Effect of dissolved oxygen concentration upon the rate of nitrification.*

- A decrease in enzymatic activity in organotrophic bacteria resulting in decreased cBOD removal efficiency
- The development of a phosphorus deficiency through the precipitation of orthophosphate with calcium ions
- The development of weak and buoyant floc particles

Because nitrifying bacteria are strict aerobes, nitrification can occur only in the presence of free molecular oxygen (O_2). Approximately 4.6 mg of oxygen are consumed per milligram of ionized ammonia oxidized to nitrate.

Insignificant nitrification occurs at dissolved oxygen concentrations lower than 0.5 mg/liter (Figure 10.5), while significant nitrification occurs at dissolved oxygen concentrations between 2 and 3 mg/liter (Table 10.8). Improved nitrification may occur at dissolved oxygen concentrations higher than 3 mg/liter, if organotrophic bacteria remove cBOD more rapidly and provide more time for nitrification.

TABLE 10.8 Dissolved Oxygen Concentration and Nitrification

Dissolved Oxygen (mg/liter)	Effect on Nitrification
<0.5	Nitrification initiated but insignificant
0.5–0.9	Rate of nitrification begins to accelerate
1.0–2.0	Rate of nitrification is significant
2.1–2.9	Sustained nitrification
3.0	Maximum rate of nitrification
>3.0	Nitrification may improve, if organotrophic bacteria remove cBOD more rapidly

TABLE 10.9 Forms of Toxicity to Nitrifying Bacteria

Form	Description or Example
Free chlorine residual	Hypochlorous acid (HOCl) or hypochlorous ion (OCl⁻)
Inorganic	Heavy metals
Organic	Phenols and recognizable, soluble cBOD
pH	<5.0
Substrate	Free ammonia or free nitrous acid
Sunlight	Ultraviolet radiation
Temperature	<5°C

TABLE 10.10 Examples of Recognizable, Soluble cBOD

Organic Compound	Formula	Number of Carbon Units
Methanol	CH_3OH	1
Methylamine	CH_3NH_2	1
Ethanol	CH_3CH_2OH	2
n-Propanol	$CH_3CH_2CH_2OH$	3
i-Propanol	$(CH_3)_2CHOH$	3
n-Butanol	$CH_3CH_2CH_2CHOH$	4
t-Butanol	$(CH_3)_3COH$	4
Ethyl acetate	$CH_3CO_2C_2H_5$	4
Aminoethanol	$CH_3NH_2CH_2OH$	2

There are several forms of toxicity to nitrifying bacteria (Table 10.9). These forms of toxicity include two unique forms for nitrifying bacteria. The unique forms of toxicity are (1) recognizable, soluble cBOD and (2) substrate.

Nitrifying bacteria are obligate autotrophs. As obligate autotrophs, they are dependent upon inorganic carbon for their carbon substrate. As obligate autotrophs, their enzymatic ability to oxidize ionized ammonia and nitrite is inhibited in the presence of several specific, short-chain (1–4 carbon units) alcohols and amines (Table 10.10). These inhibitory organic compounds are referred to as recognizable, soluble cBOD. These compounds may be discharged to the sewer system or may be produced in the sewer system or various treatment tanks under anaerobic (septic) conditions. Only when these organic compounds are either degraded to a low concentration or degraded completely does nitrification occur in the activated sludge process.

Although ionized ammonia and nitrite are the energy substrates for nitrifying bacteria, their accumulation in the activated sludge process can cause substrate toxicity to nitrifying bacteria. Substrate toxicity occurs when the ionized ammonia concentration in the aeration tank is >480 mg/liter.

With increasing pH, ionized ammonia is converted to free ammonia (Equation 10.8). The accumulation of ammonia is a function of the ionized ammonia concentration and aeration tank pH.

$$NH_4^+ + OH^- \leftrightarrow H_2O + NH_3 \qquad (10.8)$$

With decreasing pH, nitrite is converted to free nitrous acid (Equation 10.9). The accumulation of nitrous acid is a function of nitrite concentration and aeration tank pH. Substrate toxicity due to the accumulation of nitrous acid occurs because ionized ammonia is oxidized to nitrite more quickly than nitrite is oxidized to nitrate:

$$NO_2^- + H^+ \leftrightarrow HNO_2 \qquad (10.9)$$

FORMS OF NITRIFICATION

Nitrification may be complete or incomplete (Table 10.11). Because there are two groups of nitrifying bacteria and two biochemical reactions that are involved in nitrification, there are four possible forms of incomplete nitrification.

The identification of the form of nitrification that occurs in the activated sludge process is of value to an operator to (1) ensure proper nitrification, (2) provide for cost-effective operation, (3) maintain permit compliance, and (4) initiate prompt correct measures for undesired nitrification. The form of nitrification can be obtained by determining the concentrations of ionized ammonia, nitrite, and nitrate in a filtrate sample of mixed liquor effluent from an on-line aeration tank (Table 10.11).

Each form of incomplete nitrification can occur as a result of depressed temperature or a limiting process condition (Table 10.12). Limiting factors consist of (1) low dissolved oxygen concentration, (2) slug discharge of soluble cBOD, (3) swings in pH greater than ±.3 standard units, and (4) toxicity. Additional factors that may

TABLE 10.11 Forms of Nitrification

Form of Nitrification	Mixed Liquor Effluent Filtrate Concentration (mg/liter)		
	NH_4^+	NO_2^-	NO_3^-
Complete	<1	<1	>1
Incomplete #1	<1	>1	<1
Incomplete #2	>1	<1	>1
Incomplete #3	<1	>1	>1
Incomplete #4	>1	>1	>1

TABLE 10.12   Factors Responsible for Incomplete Nitrification

Form	Factor(s) Responsible
Incomplete #1	Theoretical, not likely to occur
Incomplete #2	Limiting factor
Incomplete #3	Depressed wastewater temperature
Incomplete #4	Depressed wastewater temperature and/or limiting factor

TABLE 10.13   Nitrogenous Compounds of Concern to Activated Sludge Processes

Compound	Formula	Impact
Ammonia	NH_3	Toxicity
Ionized ammonia	NH_4^+	Oxygen demand upon nitrification to NO_2^-
		Primary nitrogen nutrient for bacterial growth
		Toxicity upon conversion to NH_3
Nitrite	NO_2^-	Denitrification ("clumping") in the secondary clarifier
		Increase chlorine demand ("chlorine sponge")
		Oxygen demand upon nitrification to NO_3^-
		Toxicity
Nitrate	NO_3^-	Denitrification ("clumping") in the secondary clarifier
		Secondary nitrogen nutrient for bacterial growth
		Toxicity upon reduction to NO_2^- by *E. coli*
Organic nitrogen	TKN	Oxygen demand upon degradation
		Release of cBOD upon degradation
		Release of nBOD (NH_4^+) upon degradation

contribute to undesired nitrification include (1) decreased HRT, (2) deficiency for alkalinity, and (3) deficiency for phosphorus.

There are two forms of incomplete nitrification that result in the production and accumulation of nitrite. These forms are incomplete #3 and incomplete #4. The accumulation of nitrite is known as the "chlorine sponge," "nitrite kick," and "nitrite lock."

Nitrite reacts quickly with free chlorine and interferes with its ability to destroy coliform bacteria and pathogens in the chlorine contact tank and undesired filamentous organisms in the activated sludge process. Approximately 13 pounds of chlorine are consumed or rendered inactive for each mg/liter of nitrite accumulated per millions gallons of flow. To compensate for the chlorine sponge, the quantity of chlorine needed must be calculated and adjusted hourly, or the operational condition responsible for the accumulation of nitrite must be identified and corrected.

There are several nitrogenous compounds of concern to activated sludge processes (Table 10.13) and regulatory agencies (Table 10.14) due to their impact upon the activated sludge process and receiving waters, respectively. These compounds include ammonia, ionized ammonia, nitrite, nitrate, and organic nitrogen or total kjeldahl (TKN). Due to the adverse impact of these nitrogenous wastes upon receiving waters, many activated sludge processes are required to nitrify—that is, reduce the quantity of ammonia or ionized ammonia in the final effluent.

Nitrification in activated sludge processes can be achieved in one-stage or two-stage nitrification systems (Figure 10.6). One-stage systems consist of one aeration

TABLE 10.14 Nitrogenous Compounds of Concern to Regulatory Agencies

Compound	Formula	Impact
Ammonia	NH_3	Toxicity
Ionized ammonia	NH_4^+	Oxygen demand upon nitrification to NO_2^-
		Toxicity upon conversion to NH_3
Nitrite	NO_2^-	Oxygen demand upon nitrification to NO_3^-
		Toxicity
Nitrate	NO_3^-	Causative agent for methemoglobinemia
		Primary nitrogen nutrient for aquatic plants
		Undesired growth of aquatic plants, especially algae
Organic nitrogen	TKN	Oxygen demand upon degradation
		Release of cBOD upon degradation
		Release of nBOD (NH_4^+) upon degradation

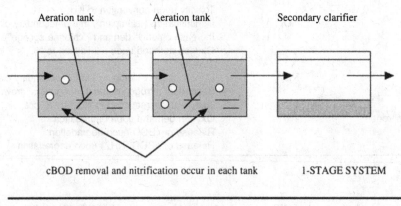

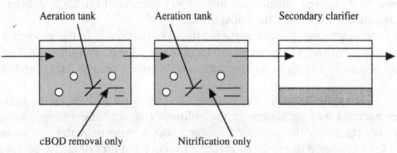

FIGURE 10.6 One-stage and two-stage nitrification systems.

tank or a series of aeration tanks that remove cBOD and nBOD. Two-stage nitrification systems consist of at least two aeration tanks or a series of aeration tanks. The first tank or series of tanks remove cBOD and the second tank or series of tanks remove nBOD (nitrify). Two-stage systems provide better process control than one-stage systems; and with regulatory requirements becoming more stringent for ammonia or ionized ammonia discharge, especially in temperate regions of the United States, two-stage nitrification systems are becoming popular.

TABLE 10.15   Benefits Obtained Through the Use of Controlled Anoxic (Denitrification) Periods

Decrease in sludge production
Destruction of undesired filamentous organisms
Improvement in process control: ensures adequate cBOD removal
Improvement in process control: ensures the presence of a "healthy" biomass
Returns some of the alkalinity to the treatment process that was lost during nitrification
Strengthens floc particles

Although many activated sludge processes are not required to nitrify, operators of these processes may promote nitrification. The presence of acceptable nitrification helps to ensure the presence of a "healthy" biomass and a satisfactory final effluent quality. The use of nitrate produced through nitrification in selected anoxic (denitrification) periods or zones provides for improved floc particle structure and decreased operational costs (Table 10.15). A comprehensive review of nitrifying bacteria is provided in *Nitrification and Denitrification in the Activated Sludge Process* in the Wastewater Microbiology Series.

11

Denitrifying Bacteria

Denitrification is the use of nitrate (NO_3^-) or nitrite (NO_2^-) by facultative anaerobic bacteria for the degradation of soluble cBOD (Equation 11.1). Unless nitrate or nitrite is present in industrial wastewaters that are discharged to an activated sludge process (Table 11.1), the activated sludge process must produce nitrate or nitrite through nitrification in order for denitrification to occur.

$$6NO_3^- + 5CH_3OH \xrightarrow{\text{facultative anaerobic bacteria}} 3N_2 + 5CO_2 + 6OH^- \qquad (11.1)$$

Denitrification requires an anoxic condition, and there are four significant criteria that must be satisfied in order to establish an anoxic condition (i.e., denitrification). The criteria are

- The presence of an abundant and active population of facultative anaerobic bacteria or denitrifying bacteria
- The presence of nitrate or nitrite
- The absence of free molecular oxygen (O_2) or the presence of an oxygen gradient (Figure 11.1)
- The presence of soluble cBOD

DENITRIFYING BACTERIA

Facultative anaerobic bacteria or denitrifying bacteria are capable of using either free molecular oxygen, nitrate, or nitrite to degrade soluble cBOD in order to obtain carbon and energy for cellular growth and activity. Although denitrifying bacteria

Wastewater Bacteria, by Michael H. Gerardi
Copyright © 2006 John Wiley & Sons, Inc.

TABLE 11.1 Industrial Wastewaters that Contain Nitrate or Nitrite

Industrial Wastewater	Nitrate (NO_3^-)	Nitrite (NO_2^-)
Corrosion inhibitor		X
Leachate (pretreated)	X	X
Meat (flavoring)	X	
Meat (preservative)		X
Meat (pretreated)	X	X
Steel	X	X

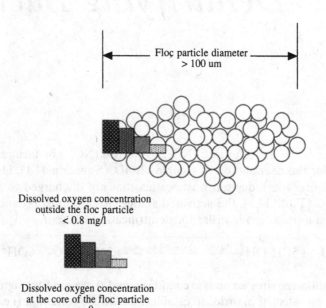

Floç particle diameter
> 100 um

Dissolved oxygen concentration
outside the floc particle
< 0.8 mg/l

Dissolved oxygen concentration
at the core of the floc particle
= 0

FIGURE 11.1 *Dissolved oxygen gradient.*

are capable of using free molecular oxygen, nitrate, and nitrite, the bacteria can only use one molecule at a time. They use the molecule that is available and provides the most carbon and energy for cellular growth and activity. Their preference always is for free molecular oxygen. When compared to the use of nitrate for the degradation of soluble cBOD, free molecular oxygen provides for more cellular growth (Equations 11.2 and 11.3) and more cellular energy (Equations 11.4 and 11.5).

$$\text{1 pound of glucose} + O_2 \xrightarrow{\text{denitrifying bacteria}} \text{0.6 pound of cells (sludge)} \quad (11.2)$$

$$\text{1 pound of glucose} + NO_3^- \xrightarrow{\text{denitrifying bacteria}} \text{0.4 pound of cells (sludge)} \quad (11.3)$$

$$\text{1 mole of glucose} + O_2 \xrightarrow{\text{denitrifying bacteria}} \text{686 kcal of energy} \quad (11.4)$$

$$\text{1 mole of glucose} + NO_3^- \xrightarrow{\text{denitrifying bacteria}} \text{636 kcal of energy} \quad (11.5)$$

Denitrifying bacteria enter activated sludge processes in fecal waste and through infiltration and inflow (I/I) as soil and water organisms. They are easily incorporated

**TABLE 11.2   Genera of Activated Sludge Bacteria that
Contain Denitrifying Species**

Achromobacter	Escherichia	Neisseria
Acinetobacter	Flavobacterium	Paracoccus
Agrobacterium	Glucononebacer	Propionibacterium
Alcaligens	Holobacterium	Pseudomonas
Bacillus	Hyphomicrobium	Rhizobium
Chromobacterium	Kingella	Rhodopseudomonas
Corynebacterium	Methanonas	Spirillum
Denitrobacillus	Moraxella	Thiobacillus
Enterobacter		Xanthomonas

into floc particles. Some denitrifying bacteria are floc formers, while most denitrifying bacteria are incorporated into floc particles through compatible surface charge or the coating action of secretions from ciliated protozoa, rotifers, and free-living nematodes.

Denitrifying bacteria are present in billions per gram of floc particle and represent approximately 80% of all bacteria flocculated and dispersed in the activated sludge process. There are numerous genera of bacteria that contain species of denitrifying bacteria (Table 11.2). The genera that contain the most species of denitrifying bacteria include *Alcaligens*, *Bacillus*, and *Pseudomonas*.

Denitrifying bacteria reproduce quickly. The generation time for most is approximately 15–30 minutes. The enzymes necessary for the use of nitrate or nitrite are formed quickly under an anoxic condition or low dissolved oxygen condition.

NITRATE AND NITRITE

Unless nitrate or nitrite is discharged to an activated sludge process in industrial wastewater, nitrate and nitrite are produced through nitrification. Although nitrite is infrequently produced through nitrification, it can accumulate in the activated sludge process during incomplete nitrification.

OXYGEN

As long as free molecular oxygen is available for bacterial use, denitrification cannot occur. Therefore, the absence of free molecular oxygen or the presence of an oxygen gradient is necessary for denitrification.

An oxygen gradient is established when the dissolved oxygen concentration is <1 mg/liter outside a floc particle that is >150 μm in diameter. As dissolved oxygen and nitrate diffuse to the core of the floc particle, denitrifying bacteria use dissolved oxygen. Once the dissolved oxygen is no longer available due to its use in degrading soluble cBOD, denitrifying bacteria begin to use nitrate. An oxygen gradient permits denitrification in the presence of measurable dissolved oxygen.

SOLUBLE cBOD

The quantity of substrate or soluble cBOD is the most important factor that influences denitrification. The greater the quantity of soluble cBOD, the greater the demand is for electron acceptors such as free molecular oxygen and nitrate. As soluble cBOD is degraded inside the bacterial cell, electrons are released from the degraded substrate. The released electrons are removed from the bacterial cell by electron acceptors. Therefore, the greater the quantity of soluble cBOD that is degraded, the greater the quantity of electron acceptors that are used. The more rapidly that oxygen is removed, the more quickly the use of nitrate (denitrification) occurs.

Denitrifying bacteria can use a large variety of soluble organic compounds as substrate including those found in domestic wastewater. Several organic compounds that are commonly added to denitrification tanks include acetate (CH_3COOH), ethanol (CH_3CH_2OH), glucose ($C_6H_{12}O_6$), and methanol (CH_3OH). Methanol usually is the organic compound of choice. Methanol is a very simplistic form of soluble cBOD. It is absorbed rapidly by bacterial cells and is degraded easily.

BIOLOGICAL REDUCTION OF NITRATE

Denitrification is one of two forms of nitrate reduction. When nitrate is reduced through bacterial activity, oxygen is removed from nitrate. Facultative anaerobic bacteria reduce nitrate to degrade soluble cBOD when free molecular oxygen is not available. This is referred to as denitrification or dissimilatory nitrate reduction, because the nitrogen in nitrate is not incorporated into cellular material; that is, nitrogen leaves the bacterial cell in molecular nitrogen (N_2) and nitrous oxide (N_2O) (Figure 11.2).

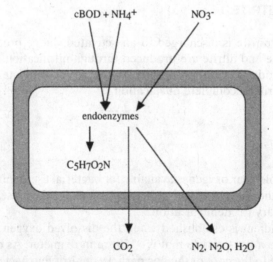

FIGURE 11.2 Dissimilatory nitrate reduction. In the absence of free molecular oxygen or presence of an oxygen gradient, facultative anaerobic bacteria remove nitrate from the bulk solution to degrade soluble cBOD. The nitrogen in the nitrate is never incorporated into new cellular material. The nitrogen in the nitrate leaves the cell as molecular nitrogen (N_2) and nitrous oxide (N_2O).

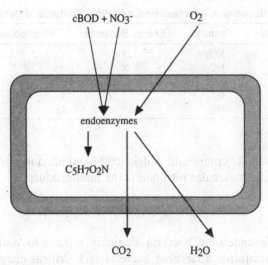

FIGURE 11.3 *Assimilatory nitrate reduction. In the absence of ionized ammonia, bacteria remove nitrate from the bulk solution for use as a nitrogen nutrient. The nitrogen in the nitrate is incorporated into new cellular material.*

While dissimilatory nitrate reduction or denitrification is used for respiratory purposes (degradation of soluble cBOD), assimilatory nitrate reduction is used to provide bacterial cells with the nitrogen nutrient (Figure 11.3). When present, ionized ammonia (NH_4^+) is used as the nitrogen nutrient by bacterial cells for the synthesis (assimilation) of cellular material. In order for bacterial cells to used nitrogen, nitrogen must be in the −3 valence or oxidation state. In ionized ammonia, nitrogen exists in the −3 oxidation state and is readily available for bacterial use; that is, it is preferred.

When ionized ammonia is no longer available, nitrate (NO_3^-) is used as the nitrogen nutrient. However, nitrogen exists in the +5 oxidation state in nitrate. Therefore, when nitrate is used as the nitrogen nutrient, oxygen is removed from nitrate inside the bacterial cell and hydrogen is added to the nitrogen. By removing oxygen and adding hydrogen, the nitrogen is reduced to a −3 oxidation state, and ionized ammonia is formed. The nitrogen in the newly formed ionized ammonia is then incorporated (assimilated) into new cellular material; that is, nitrogen does not leave the cell. This form of nitrate reduction is referred to as assimilatory nitrate reduction.

Dissimilatory nitrate reduction and assimilatory nitrate reduction can occur simultaneously in an activated sludge process. In the presence of soluble cBOD and nitrate ions and the absence of free molecular oxygen and ionized ammonia, nitrate is used to degrade soluble cBOD (denitrification) and provide the nitrogen nutrient.

Denitrification proceeds in a step-by-step reduction from nitrate to molecular nitrogen (Equation 11.6). There are five nitrogenous molecules and four biochemical steps involved in denitrification. The molecules are nitrate (NO_3^-), nitrite (NO_2^-), nitric oxide (NO), nitrous oxide (N_2O), and molecular nitrogen (N_2). Nitrate always is an initial energy substrate (Table 11.3), while nitrite may be an initial sub-

TABLE 11.3 Energy Substrates, Intermediates, and Final Products of Denitrification

Nitrogenous Compound	Formula	Energy Substrate	Intermediate	Final Product
Nitrate ion	NO_3^-	X		
Nitrite ion	NO_2^-	X	X	
Nitric oxide	NO		X	
Nitrous oxide	N_2O		X	
Molecular nitrogen	N_2			X

strate or an intermediate compound. Nitric oxide and nitrous oxide are intermediate compounds, while molecular nitrogen is the final product.

$$NO_3^- \rightarrow NO_2^- \rightarrow NO \rightarrow N_2O \rightarrow N_2 \tag{11.6}$$

Most facultative anaerobic bacteria denitrify nitrate to molecular nitrogen. However, some facultative anaerobic bacteria lack critical enzymes to denitrify nitrate to molecular nitrogen. For example, some bacteria such as *Escherichia coli* denitrify nitrate to nitrite and stop. In addition, adverse operational conditions also permit the production of intermediate compounds by facultative anaerobic bacteria.

Although there are four biochemical steps involved in denitrification, there are only two energy-yielding steps. These reactions are the reduction of nitrate (Equation 11.7) and the reduction of nitrite (Equation 11.8). These two reactions can be combined and presented as an overall energy yielding reaction (Equation 11.9).

$$6NO_3^- + 2CH_3OH \xrightarrow{\text{denitrifying bacteria}} 6NO_2^- + 2CO_2 + 4H_2O \tag{11.7}$$

$$6NO_2^- + 3CH_3OH \xrightarrow{\text{denitrifying bacteria}} 3N_2 + 3CO_2 + 6OH^- \tag{11.8}$$

$$6NO_3^- + 5CH_3OH \xrightarrow{\text{denitrifying bacteria}} 3N_2 + 5CO_2 + 7H_2O + 6OH^- \tag{11.9}$$

The overall energy yielding reaction produces alkalinity in the form of hydroxyl ions (OH^-) and bicarbonate ions (HCO_3^-). Bicarbonate ions are produced when carbon dioxide dissolves in the wastewater to form carbonic acid (H_2CO_3), which dissociates to form bicarbonate ions. Approximately 50% of the alkalinity lost during nitrification is returned to the process through denitrification.

Much of the energy obtained from the degradation of cBOD is used for cellular synthesis—that is, the production of new bacterial cells ($C_5H_7O_2N$) or sludge (Equation 11.10). Because less carbon from the cBOD is assimilated into new bacterial cells when nitrate is used to degrade the cBOD as compared to the use of free molecular oxygen, more carbon dioxide is produced when cBOD is degraded with nitrate. Much of the carbon dioxide that is produced does not dissolve in the wastewater.

$$\text{Nitrate} + \text{cBOD} \xrightarrow{\text{denitrifying bacteria}} \text{cells (sludge)} + \text{water} + \text{carbon dioxide} \tag{11.10}$$

Although five gases are produced during denitrification, only three gases escape to the atmosphere from the wastewater (Table 11.4). The majority of gases that are

TABLE 11.4   Fate of Gases Produced Through Denitrification

Gas	Formula	Fate
Molecular nitrogen	N_2	Insoluble in wastewater Leaves as escaping bubbles
Carbon dioxide	CO_2	Although soluble in wastewater, forms carbonic acid/bicarbonate alkalinity Some leaves as escaping bubbles
Nitrous oxide	N_2O	Insoluble in wastewater Leaves as escaping bubbles
Ammonia	NH_3	Converted to NH_4^+ at pH values <9.4 NH_4^+ dissolves in wastewater
Nitric oxide	NO	Not usually released from bacterial cell Does not accumulate

produced consists of molecular nitrogen and carbon dioxide. Often, these gases alone with nitrous oxide become entrapped in floc particles and contribute to settleability problems in the secondary clarifier and thickener before the gases escape to the atmosphere.

Denitrification occurs between the oxidation–reduction potential (ORP) or redox values of +100 mV and –100 mV and is influenced by three significant operational factors. These factors are pH, temperature, and nutrients. The optimal pH range for denitrification is from 7.0 to 7.5, while depressed activity for facultative anaerobic bacteria occurs at pH values <6.0 and >8.0.

Denitrification is biologically mediated and occurs more rapidly with increasing wastewater temperature. Additionally, increasing wastewater temperature results in less wastewater affinity for dissolved oxygen, which contributes to more rapid denitrification. Denitrification is inhibited at wastewater temperatures <5°C.

Nitrogen and phosphorus are critical major nutrients for the degradation of soluble cBOD and the growth of bacterial cells (sludge). Adequate quantities of these nutrients for the degradation of soluble cBOD with free molecular oxygen are considered to be present when the mixed liquor effluent of an aeration tank contains these residual quantities for nitrogen and phosphorus:

- 1.0 mg/liter NH_4^+ or 3.0 mg/liter NO_3^- for nitrogen
- 0.5 mg/liter $H_2PO_4^-$ or HPO_4^{2-}

Since the degradation of soluble cBOD with nitrate produces less bacterial growth than with free molecular oxygen, these values for nitrogen and phosphorus are adequate for the effluent filtrate from a denitrifying tank.

Denitrification occurs whenever an anoxic condition exists. The anoxic condition can occur by design or accident. Designed anoxic conditions consist of (1) the use of a denitrification tank to satisfy a total nitrogen discharge limit and (2) the use of an anoxic period or zone to improve treatment plant performance. Accidental anoxic conditions are most commonly observed in secondary clarifiers.

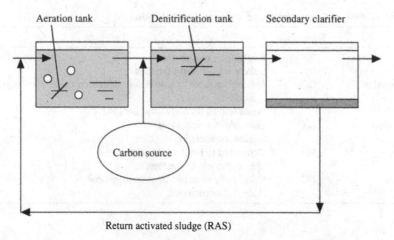

FIGURE 11.4   Denitrification tank.

DENITRIFICATION TANK

A denitrification tank (Figure 11.4) is used with an aeration tank or series of aeration tanks to satisfy a total nitrogen discharge limit. In the aeration tank, ammonification occurs. Ammonification is the release of ionized ammonia from organic nitrogen compounds. The released ionized ammonia and the raw influent ionized ammonia that enter the aeration tank undergo nitrification resulting in the production of nitrate.

The nitrate, bacteria, and residual dissolved oxygen in the aeration tanks are discharged to the denitrification tank. Here, slow subsurface mixing action is provided to suspend the bacteria, and a carbon source (cBOD) such as methanol (CH_3OH) is added. The suspended bacteria rapidly remove the residual dissolved oxygen as they degrade the methanol. Once the dissolved oxygen is removed, the facultative anaerobic bacteria continue to degrade the methanol with nitrate.

Because nitrate is reduced in the denitrification tank, the nitrogen within the nitrate escapes to the atmosphere as molecular nitrogen and nitrous oxide. The nitrogen is not discharged to the clarifier or receiving body of water.

Caution should be used when adding a carbon source (cBOD) to a denitrification tank. If too much of the carbon source is added, some may leave the plant in the final effluent. This represents an increase in the effluent cBOD and a loss of money. Therefore, it is necessary to know how much carbon should be added to denitrify. For methanol, 2.47 mg/liter of methanol are required for each mg/liter of nitrate present in order to denitrify completely. Often a carbon-to-nitrate ion ratio of 3 : 1 is used as a guideline. Most denitrifying tanks have a hydraulic retention time of 30–60 minutes.

USE OF AN ANOXIC PERIOD OR ZONE

An anoxic period or zone may be used to improve treatment plant performance. The use of an anoxic period or zone of 1–2 hours on an as-needed basis or as part of the standard operating procedures of the treatment process can provide the following benefits for an activated sludge process:

- Control of undesired filamentous organism growth
- Improvement in floc formation
- Decrease in sludge production

Some filamentous organisms such as *Haliscomenobacter hydrossis*, Nocardioforms, *Sphaerotilus natans*, and type 1701 are strict aerobes. By exposing strict aerobic filamentous organisms to anoxic periods of 1–2 hours, their growth is controlled, because these organisms can only use free molecular oxygen for cellular activity and degradation of soluble cBOD.

Floc particles contain strict aerobic bacteria and facultative anaerobic bacteria. Several of each bacteria are floc-forming organisms. By exposing these organisms to anoxic periods, the number of strict aerobes is gradually reduced, while the number of facultative anaerobes is gradually increased. Because floc particles become denser and firmer with an increasing percentage of facultative anaerobic bacteria, the use of anoxic periods produces floc particles that settle better and are more tolerant of shearing action.

Because the degradation of soluble cBOD with nitrate produces less sludge as compared to the degradation of soluble cBOD with free molecular oxygen, the use of anoxic periods results in decreased sludge production. Decreased sludge production results in the use of less polymers and coagulants for thickening and dewatering purposes providing an overall decrease in sludge disposal costs.

Controlled anoxic periods can be established by the following operational measures:

- Terminating the addition of air to an aeration tank and permitting the use of nitrate in the aeration tank for 1–2 hours provided the settled solids do not adversely affect the aeration equipment during restart
- Establishing an anoxic tank in the first aeration tank in plug-flow mode of operation (Figure 11.5)
- Establishing an anoxic zone in an aeration tank with a surface aerator (Figure 11.6) or an oxidation ditch (Figure 11.7)

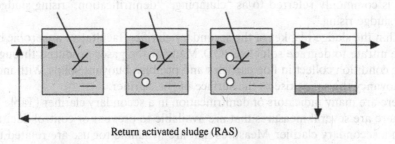

Return activated sludge (RAS)

FIGURE 11.5 *Plug-flow mode of operation with an anoxic tank.*

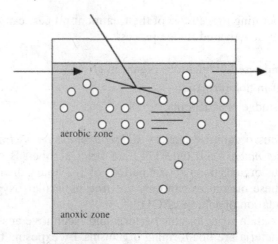

FIGURE 11.6 Anoxic zone in a surface aerated tank.

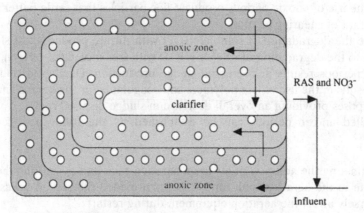

FIGURE 11.7 Anoxic zones in an oxidation ditch.

ACCIDENTAL ANOXIC CONDITION

An accidental anoxic condition or undesired anoxic condition is most often associated with poor settling solids and loss of solids in the secondary clarifier. This condition is commonly referred to as "clumping," "denitrification," rising sludge," or "dark sludge rising."

Within the sludge blanket of the secondary clarifier, facultative anaerobic bacteria use nitrate to degrade soluble cBOD. Many of the gases produced through the anoxic condition collect in floc particles and produce buoyant solids. With increasing buoyancy the solids rise to the surface of the clarifier.

There are many indicators of denitrification in a secondary clarifier (Table 11.5), and there are several measures that are available to prevent or control denitrification in a secondary clarifier. Measures that are available for use are related to the presence or absence of a nitrification requirement (ammonia discharge limit).

TABLE 11.5 Indicators of Denitrification

Indicator	Rationale
Bubbles	Release of N_2, CO_2, and N_2O during anoxic degradation of soluble cBOD
Bubbles and solids	Bubbles entrapped in floc particles
Dark solids	Old (dark) sludge present in the treatment system in order to grow nitrifying bacteria to produce NO_3^-
Decrease in NO_3^- across secondary clarifier	NO_3^- used for anoxic degradation of soluble cBOD
Escape of N_2O	Capture and detection of N_2O over secondary clarifier
Increase in alkalinity and pH	Alkalinity produced during denitrification
Reduction in redox (ORP) potential	Decrease in NO_3^- results in decrease in ORP
Rising and floating solids	Presence of buoyant floc particles (entrapped gases)

If an activated sludge process is not required to nitrify, then denitrification in the secondary clarifier can be corrected by terminating nitrification. Measures available to terminate nitrification include the following:

- Reducing dissolved oxygen concentration in the aeration tank
- Reducing MCRT
- Reducing MLVSS
- Reducing the number of aeration tanks that are used in the treatment process

If an activated sludge process is required to nitrify, then denitrification in the secondary clarifier can be corrected by the following measures:

- Increasing the RAS rate
- Increasing the RAS rate and adding coagulants and/or polymers to thicken the secondary solids
- Using anoxic periods with the aeration tanks
- Using plug-flow mode of operation and using an aeration tank as an anoxic tank

A comprehensive review of denitrifying bacteria is provided in *Nitrification and Denitrification in the Activated Sludge Process* in the Wastewater Microbiology Series.

12

Poly-P Bacteria

Phosphorus (P) is a major nutrient that is necessary to all living cells. It is an essential element in the production of adenosine triphosphate or ATP (Figure 12.1), the nucleic acids DNA and RNA, phospholipids, teichoic acids, and teichuronic acids.

ATP serves as a high-energy molecule and is used in the transfer of energy within the cell. Phospholipids are key components in the structure of cell membranes, while teichoic acids and teichuronic acids are key components in the structure of cell walls of Gram-positive bacteria. Phosphorus also is stored in cells as intracellular volutin granules or polyphosphates.

Phosphorus may be 1–3% of the dry weight of a bacterium. Although the phosphorus content is approximately one-fifth of the nitrogen content of the bacterium, the actual phosphorus content may vary from one-seventh to one-third of the nitrogen content depending upon environmental conditions.

Phosphorus is a nutrient required in the growth of aquatic plants. Often phosphorus is the limiting nutrient—that is, the concentration of phosphorus in waters determines the quantity of vegetative growth. Therefore, the introduction of trace amounts of phosphorus into receiving waters can have profound and undesired effects on the quality of the receiving waters.

The undesired growth of algae often is triggered at orthophosphate concentrations as low as 0.5 mg/liter. The presence of algal blooms as well as phytoplankton results in a rapid and significant deterioration in water quality. Phosphorus pollution of natural waters is mainly responsible for eutrophication and occurs chiefly as a result of phosphorous-rich effluents from wastewater treatment plants.

Several environmental problems are associated with the rapid growth of aquatic plants. These problems include clogging of the receiving waters as well as color, odor, taste, and turbidity issue, if the receiving waters are used as potable water

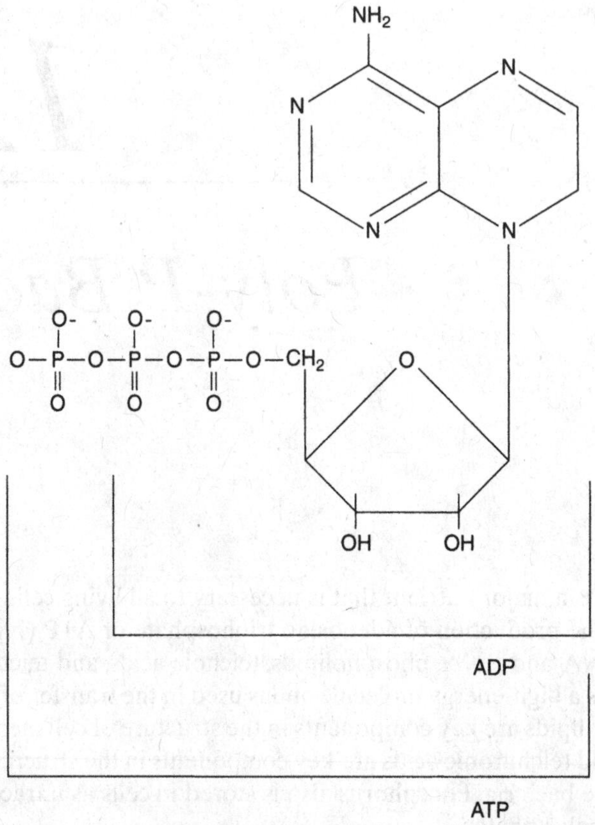

FIGURE 12.1 ATP. ATP or adenosine triphosphate contain three phosphate groups and two high-energy phosphate bonds. When bacterial cells need energy, one high-energy phosphate bond is broken to release energy, and ADP or adenosine diphosphate is formed. When bacterial cells stored energy, a phosphate group is added to ADP through the production of a high-energy phosphate bond.

sources. Additionally, and more importantly, the die-off of large numbers of aquatic plants contributes to oxygen depletion and eutrophication. Oxygen is removed from the receiving waters by bacteria and other organisms as they decompose the dead plants. Eutrophication or rapid aging of the receiving waters occurs as the nonde-composable portions of the dead plants accumulate in the receiving waters. The rapid growth of aquatic plants also lowers the value of the receiving waters for fishing, industrial use, and recreational use.

Most often phosphorus is found in wastewater in quantities greater than those required for the growth of aquatic plants. Therefore, in order to prevent or reduce phosphorous-related water quality problems, state and federal regulatory agencies often require phosphorus removal at wastewater treatment plants. Because phosphorus reacts quickly with minerals such as aluminum, calcium, and iron, little phosphorus leaches from the soil. Also, little phosphorus leaches from the soil when it is applied to the soil as a fertilizer.

The requirement for phosphorus removal is becoming more common for municipal and industrial wastewater treatment plants. Discharge limits for total phosphorus at these plants often are $\leq 2\,\text{mg/liter}$.

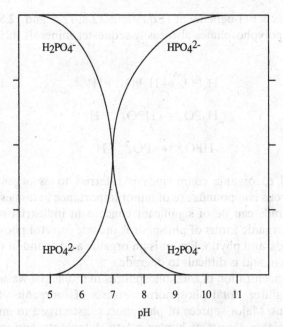

FIGURE 12.2 Distribution of $H_2PO_4^-$ and HPO_4^{2-} in the mixed liquor.

Phosphorus exists in inorganic and organic forms. Inorganic forms of phosphorus include orthophosphates and polyphosphates. Orthophosphates are available for biological metabolism without further breakdown and are considered to be the readily available nutrients for phosphorus for bacterial use in wastewater treatment plants and aquatic plants in natural waters.

Orthophosphates include PO_4^{3-}, HPO_4^{2-}, $H_2PO_4^-$, and H_3PO_4. The most common forms of orthophosphate in wastewater treatment plants are HPO_4^{2-} and H_2PO^-. The relative quantity of each form is pH-dependent (Figure 12.2). The form of orthophosphate present is produced through dissociation (Equation 12.1). Within the pH operating range of most wastewater treatment plants, HPO_4^{2-} is dominant at pH values greater than 7, while $H_2PO_4^-$ is dominant at pH values greater than 7.

$$H_2PO_4^- \leftrightarrow HPO_4^{2-} + H^+ \tag{12.1}$$

Polyphosphates are complex molecules with two or more phosphorous atoms, oxygen atoms, and perhaps hydrogen atoms. Polyphosphates are represented by the chemical formula for the pyrophosphate ion ($P_2O_7^{3-}$). Pyrophosphate is the first in a series of unbranched-chain polyphosphates (i.e., $P_2O_7^{3-}$, $P_3O_{10}^{5-}$, ...). Polyphosphates undergo hydrolysis very slowly and release orthophosphate (Equation 12.2). Hydrolysis can be chemically mediated or biologically mediated by bacteria and algae.

$$H_2P_2O_7 + H_2O \rightarrow 2H_3PO_4 \tag{12.2}$$

Hydrolysis of polyphosphates is influenced by many factors including retention time and pH in an aeration tank. The principle form of orthophosphate obtained from the hydrolysis is pH-dependent (Equations 12.3, 12.4, and 12.5). Due to their stability in water, polyphosphates also easily sequester minerals such as aluminum, calcium, and iron.

$$H_3PO_4 \leftrightarrow H_2PO_4^- + H^+ \tag{12.3}$$

$$H_2PO_4^- \leftrightarrow HPO_4^{2-} + H^+ \tag{12.4}$$

$$HPO_4^{2-} \leftrightarrow PO_4^{3-} + H^+ \tag{12.5}$$

Phosphate tied to organic compounds is referred to as organic phosphorus. Organic phosphorous compounds are of minor importance in domestic wastewater, but these compounds can be of significant concern in industrial wastewater and sludge. Common organic forms of phosphorus include inositol phosphates, nucleic acids, phospholipids, and phytin. Phytin is an organic acid found in vegetables such as corn and soybean and is difficult to degrade.

The average concentration of total phosphorus in municipal wastewater is in the range of 10–20 mg/liter. Total phosphorus consists of inorganic phosphorus and organic phosphorus. Major sources of phosphorus discharged to municipal waste-water treatment plants consist of human waste, detergents, and industrial waste. Orthophosphate makes up approximately 50–70% of the total phosphorus, while polyphosphates and organic phosphorus make up the remaining 30–50% of the total phosphorus.

When orthophosphate, polyphosphate, and organic phosphorous compounds enter an activated sludge process these compounds undergo biological and chemically changes and experience several fates (Figure 12.3). Some organic phosphorous compounds are removed from the wastewater when particulate, organic phosphorous compounds or phosphorous compounds adsorbed to solids settle out in the primary clarifier.

In the activated sludge process, phosphorous compounds undergo several fates. With sufficient hydraulic retention time (HRT), organic phosphorous compounds are degraded through microbial activity, and orthophosphate is released in the aeration tank. With sufficient HRT, polyphosphates are biologically and chemically hydrolyzed, and orthophosphate is released in the aeration tank. Principal organisms responsible for the mineralization or degradation of phosphorous compounds include actinomycetes such as *Streptomyces*, bacteria such as *Arthrobacter* and *Bacillus*, and fungi such as *Aspergillus* and *Penicillium*. These organisms produce phosphatase, an enzyme that releases orthophosphate from phosphorus-containing compounds.

Orthophosphate is the readily available phosphorous nutrient for bacterial growth and energy transfer. As a readily available nutrient, phosphorus is removed from the bulk solution from the aeration tank and incorporated or assimilated into cellular material as bacteria degrade substrate (soluble cBOD) and reproduce (sludge production). Here, assimilated phosphorus makes up 1–3% of the bacterial weight (mixed liquor volatile suspended solids).

If a deficiency for orthophosphate occurs in the activated sludge process, the production of nutrient-deficient floc particles or sludge and the undesired and excessive growth of nutrient-deficient filamentous organisms may occur.

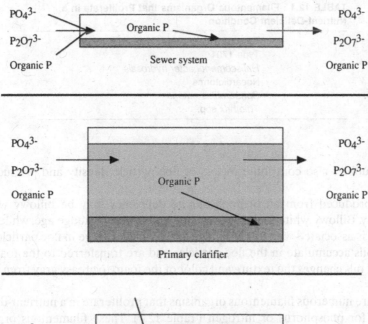

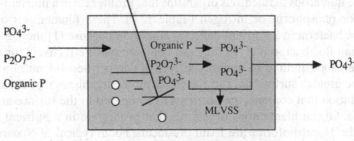

FIGURE 12.3 *Movement of phosphorus in the activated sludge process. Phosphorus enters the sewer system in the inorganic form (phosphate and polyphosphate) and organic form. In the biofilm and sediment of the sewer system, some polyphosphate is hydrolyzed to orthophosphate and some organic phosphorus is degraded to release phosphate. In the primary clarifier, some organic phosphorus is removed in the sludge blanket when organic phosphorus compounds settled out. In the aeration tank, phosphate is removed by bacteria as the phosphorus nutrient and incorporated into new cells (MLVSS). Some polyphosphate is hydrolyzed to form orthophosphate, and some organic phosphorus is degraded to release phosphate.*

Nutrient-deficient floc particles and nutrient-deficient filamentous organisms adversely affect solids settleability in the secondary clarifier and may be responsible for foam production and accumulation.

During a nutrient deficiency for orthophosphate (<0.05 mg/liter), soluble cBOD is absorbed by bacterial cells in floc particles. However, the soluble cBOD cannot be degraded due to the lack of adequate phosphorus. Therefore, the cBOD is converted to an insoluble polysaccharide (starch) and stored in the floc particle, until orthophosphate is available for its degradation. The stored polysaccharide is less dense than water, and its storage between bacterial cells results in a loss of floc particle density. The polysaccharides also capture air and gas bubbles. The captured air

TABLE 12.1   Filamentous Organisms that Proliferate in a
Nutrient-Deficient Condition

Type 021N
Type 1701
Haliscomenobacter hydrossis
Nocardioforms
Sphaerotilus natans
Thiothrix spp.

and gas bubbles also contribute to loss of floc particle density and production of foam.

Foam produced from an orthophosphate deficiency may be billowy white or greasy gray. Billowy white foam is associated with a young sludge age, while greasy gray foam is associated with an old sludge age. As bacteria age in floc particles, their secreted oils accumulate in the floc particles and are transferred to the foam. This transfer of oils changes the texture and color of the foam to greasy gray from billowy white.

There are numerous filamentous organisms that proliferate in a nutrient-deficient condition for phosphorus or nitrogen (Table 12.1). These filamentous organisms outgrow floc bacteria in a nutrient deficient condition because (1) they require less nutrients than floc bacteria or (2) they can compete more effectively for nutrients when nutrients are limited in quantity. Effective competition for nutrients is provided by the greater surface area of the filamentous organisms that is exposed to the bulk solution that contains the nutrients as compared to the surface area of the floc bacteria. Of the filamentous organisms that proliferate in a nutrient deficient condition, the Nocardioforms are foam producers. Foam typical of Nocardioforms is viscous and chocolate brown.

In the aeration tank, orthophosphate may be incorporated into floc particles as insoluble hydroxyapatite ($CaOH(PO_4)_3$). This occurs naturally without chemical addition. If the dissolved oxygen concentration of the aeration tank is relatively low and much of the carbon dioxide released from the degradation of soluble cBOD remains in solution (i.e., it is not stripped to the atmosphere), the pH of the aeration tank decreases. The decrease occurs because carbon dioxide dissolves in the mixed liquor and carbonic acid (H_2CO_3) is produced. Under this condition, orthophosphate remains in solution as the $H_2PO_4^-$ ion.

However, if the dissolved oxygen concentration of the aeration tank is relatively high and much of the carbon dioxide in the aeration tank is stripped to the atmosphere, little carbonic acid is produced and the pH of the aeration tank increases. Under this condition, orthophosphate is present as the HPO_4^{2-} ion. If this occurs in hard water (containing calcium as Ca^{2+}), orthophosphate is precipitated from solution as hydroxyapatite and incorporated into floc particles (Equation 12.4).

$$5Ca^{2+} + 3HPO_4^{2-} + H_2O \rightarrow CaOH(PO_4)_3 + 4H^+ \tag{12.6}$$

Orthophosphate may remain in solution in the aeration tank in two forms. It may remain in solution in ionic form as determined by pH, or it may remain in solution sequestered (bonded in solution) to an alkali metal.

TABLE 12.2 Biological and Chemical Measures Available for Phosphorus Removal

Measure	Description
Assimilation	Incorporation of phosphorus as a nutrient used in cellular synthesis and energy transfer. Phosphorus makes up 1–3% of bacterial weight or sludge.
Enhanced biological phosphorus removal	Incorporation of phosphorus as a nutrient used in cellular synthesis, energy transfer, and polyphosphate granules. Phosphorus makes up 6–7% of bacterial weight or sludge.
Hydroxyapatite production	Production of insoluble hydroxyapatite ($CaOH(PO_4)_3$) during low dissolved oxygen concentration and increasing pH in the aeration tank.
Chemical precipitation	Use of alum, ferric chloride, ferrous sulfate, or lime to precipitate orthophosphate as a metal salt.
Biological-mediated/chemical precipitation	Chemical precipitation of orthophosphate released by bacteria from enhanced biological phosphorus removal measure.

Effluent phosphorus from the activated sludge process is approximately 90% orthophosphate. The orthophosphate may be as soluble ions or sequestered orthophosphate. To reduce the concentration of effluent phosphorus from an activated sludge process, an advanced wastewater treatment measure is required. Advanced wastewater treatment consists of those biological, chemical, and physical measures that remove

- Inorganic and organic suspended solids
- Phosphorus-containing and nitrogen-containing compounds that contribute to eutrophication
- Slowly degradable or nondegradable organic compounds

Phosphorus can be removed in municipal wastewater treatment plants through biological and chemical treatment measures (Table 12.2). Several of these measures are considered to be advanced wastewater treatment measures and include chemical precipitation of phosphorus, enhanced biological phosphorus removal (EBPR), and biological-mediated/chemical precipitation of phosphorus.

ENHANCED BIOLOGICAL PHOSPHORUS REMOVAL

Enhanced biological phosphorus removal (EBPR) or "luxury uptake of phosphorus" occurs when phosphorus uptake by bacteria is in excess of cellular requirements. Typically, activated sludge phosphorus content is approximately 1–3%, while the activated sludge phosphorus content is approximately 6–7% when EBPR is used.

EBPR is relatively inexpensive and is capable of removing phosphorus to low effluent concentrations. EBPR also reduces chemical costs and sludge disposal costs that are associated with chemical precipitation of phosphorus.

EBPR incorporates the use of two groups of bacteria, fermentative bacteria and poly-P bacteria. Poly-P bacteria are known also as phosphorus accumulating organisms (PAO). At least two treatment tanks, an anaerobic (fermentative) tank and an

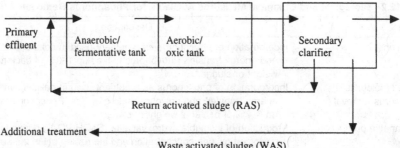

FIGURE 12.4 *Enhanced biological phosphorus removal system. In the anaerobic zone, soluble cBOD from the primary clarifier is fermented in the absence of free molecular oxygen and nitrate. The fermentation of soluble cBOD results in the production of a variety of volatile fatty acids. The acids are rapidly absorbed by the poly-P bacteria. The absorption of fatty acids results in the production of stored food (insoluble starch granules) and the release of orthophosphate by the poly-P bacteria. In the aerobic zone, orthophosphate released by the poly-P bacteria and orthophosphate in the primary clarifier effluent are absorbed by the poly-P bacteria as the starch granules are solublized and degraded. Due to the different bacterial events in the anaerobic zone and aerobic zones, the bacteria/sludge contains approximately 6–7% phosphorus on a dry weight basis as compared to 1–3% phosphorus in typical activated sludge. The sludge in the aerobic zone that contains the elevated concentration of phosphorus is discharged to the secondary clarifier where it is either wasted or returned to the anaerobic zone.*

TABLE 12.3 Significant Fermentative Bacteria in the Anaerobic (Fermentative) Tank

Aeromonas
Citrobacter
Klebsiella
Pasteurella
Proteus
Serratia

TABLE 12.4 Significant Poly-P Bacteria in the Aerobic Tank

Achromobacter	*Escherichia*
Acinobacter	*Klebsiella*
Aerobacter	*Microlunatus*
Aeromonas	*Moraxella*
Arthrobacter	*Mycobacterium*
Bacillus	*Pasteurella*
Beggiatoa	*Proteobacter*
Citrobacter	*Pseudomonas*
Enterobacter	*Vibrio*

aerobic tank, are used for EBPR (Figure 12.4). The fermentative bacteria are facultative anaerobes and anaerobes, while the poly-P bacteria with exception are strict aerobes. The fermentative bacteria (Table 12.3) and poly-P bacteria (Table 12.4) enter the EBPR process as fecal bacteria and soil and water bacteria from inflow and infiltration (I/I). The key to EBPR is the exposure of poly-P bacteria to alternating anaerobic and aerobic conditions.

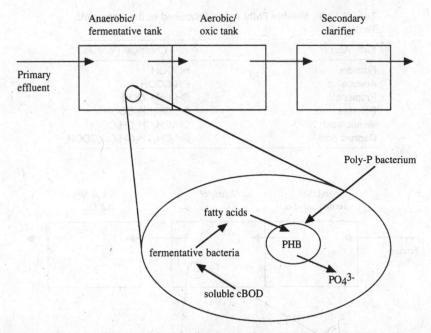

FIGURE 12.5 *Bacterial activity in the anaerobic tank in EBPR system. In the anaerobic tank, soluble cBOD is fermented in the absence of free molecular oxygen and nitrate. The fermentation process produces a variety of volatile fatty acids. The acids are rapidly absorbed by poly-P bacteria and stored as insoluble starch granules (PHB). In order to absorb the fatty acids and covert the fatty acids to PHB, energy in the form of orthophosphate is lost to the bulk solution.*

In the anaerobic tank (Figure 12.5), fatty acids (particularly acetate and propionate) are produced through the anaerobic activity of fermentative bacteria. These compounds serve as substrate for the proliferation of aerobic poly-P bacteria that are in the anaerobic tank. However, the poly-P bacteria cannot utilize (degrade) the fermentative compounds in an anaerobic condition. The degradation of these compounds by poly-P bacteria occurs only in the presence of free molecular oxygen (aerobic tank) or nitrate (NO_3^-).

Free molecular oxygen and nitrate ions should be absence in the anaerobic (fermentative) tank. A residual quantity of either free molecular oxygen or nitrate is quickly exhausted in the anaerobic tank, but requires a longer retention time and more soluble cBOD in the anaerobic tank in order to exhaust the free molecular oxygen or nitrate ions. The presence of free molecular oxygen or nitrate in the anaerobic tank interferes with the phosphorus removing ability of the EBPR process. Interference occurs as a result of an increase in redox potential and inability to produce fatty acids that are necessary for the release of phosphorus.

Fermentation is the microbial degradation of soluble organic compounds (cBOD) without the use of free molecular oxygen or nitrate. Significant fermentative organic compounds or substrate produced in the anaerobic tank include alcohols and a variety of soluble fatty acids (Table 12.5).

In the anaerobic tank, poly-P bacteria rapidly absorb the fatty acids and polymerize (store) the acids as an insoluble starch (poly-β-hydroxybutyrate or PHB). PHB in poly-P bacteria serves two important functions. First, it helps the bacteria

TABLE 12.5 Soluble Fatty Acids Produced in the Anaerobic Tank

Fatty Acid	Chemical Formula
Formate	$HCOOH$
Acetate	CH_3COOH
Propionate	CH_3CH_2COOH
Butyrate	$CH_3CH_2CH_2COOH$
Valeric acid	$CH_3CH_2CH_2CH_2COOH$
Caproic acid	$CH_3CH_2CH_2CH_2CH_2COOH$

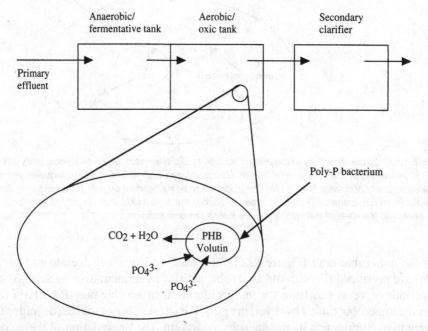

FIGURE 12.6 *Bacterial activity in the aerobic tank in EBPR system. In the aerobic tank, PHB is solublized and degraded in the presence of free molecular oxygen. The degradation of PHB results in the production of carbon dioxide, water, and new cellular material (MLVSS). Phosphate released in the anaerobic tank as well as phosphorus in the primary clarifier effluent are absorbed by the poly-P bacteria and stored as volutin.*

to grow and rebuild polyphosphates by taking up soluble phosphate. Second, PHB along with polyphosphates help aerobic poly-P bacteria to survive in an anaerobic condition.

The polymerization of the fatty acids requires an expenditure of cellular energy by the poly-P bacteria. This expenditure or energy is the breakdown or release of orthophosphate from the poly-P bacteria to the bulk solution of the anaerobic tank. As a result of the release of orthophosphate from the poly-P bacteria, the anaerobic tank contains two pools of phosphorus, namely, the phosphorus in the influent wastewater (feed phosphorus) and the released phosphorus by the poly-P bacteria.

In the aerobic tank (Figure 12.6) the poly-P bacteria use free molecular oxygen to degrade the stored PHB as a carbon and energy source. Concurrently, the poly-

TABLE 12.6 Nutrient Removal Processes

Process Name	Nutrient Removed		Phosphorus Removal		Chemical Precipitation
	Nitrogen	Phosphorus	Mainstream	Sidestream	
A/O		X	X		
Phostrip		X		X	X
A²/O	X	X	X		
Bardenpho	X	X	X		
UCT	X	X	X		

P bacteria absorb orthophosphorus in order to store the energy released from the degraded PHB. The poly-P bacteria obtain so much energy from the degraded PHB that they absorb not only the released phosphorus but also large quantities of feed phosphorus. The absorbed phosphorus is assimilated into macromolecules and stored as polyphosphate granules or volutin. Phosphorus removal is achieved when the bacteria (sludge) are wasted from the secondary clarifier. Sludge that is not wasted is returned to the anaerobic tank where the EBPR process is repeated. By exposing the poly-P bacteria to alternating anaerobic and aerobic conditions, the poly-P bacteria are stressed and take up phosphorus in excess of normal cellular requirements.

There are several processes available for EBPR. Alternating exposure of poly-P bacteria to anaerobic and aerobic tanks can be accomplished in the main biological treatment process ("mainstream") or in the return sludge ("sidestream") (Table 12.6). A mainstream process for biological phosphorus removal contains an anaerobic tank along the main liquid process stream from influent to effluent. A side stream process contains an anaerobic tank that is aside of the main liquid process steam.

All nutrient removal processes for phosphorus remove excess phosphorus biologically, except the Phostrip process that incorporates chemical precipitation of phosphorus. The removed phosphorus from these processes is found biologically in the bacterial cells or chemically in a precipitate within the sludge. When biological phosphorus removal is combined with nitrification and denitrification for nitrogen removal, the removal of phosphorus and nitrogen is known as biological nutrient removal (BNR) or combined phosphorus/nitrogen removal.

Nitrification is the biological oxidation of ionized ammonia (NH_4^+) to nitrate (NO_3^-). Denitrification is the biological used of nitrate to degrade soluble cBOD in the absence of free molecular oxygen. When nitrate is used to degrade soluble cBOD, the nitrogen in the nitrate leaves the wastewater and is returned to the atmosphere as molecular nitrogen (N_2) and nitrous oxide (N_2O).

There are two EBPR processes that are available in the United States that remove phosphorus only. These processes are the A/O (Figure 12.7) and the Phostrip (Figure 12.8). The A/O (aerobic/oxic) process is patented by Air Products and Chemicals, Inc. The A/O process is a mainstream process.

The Phostrip process is a sidestream process and includes biological and chemical measures for phosphorus removal. The Phostrip process has a stripper tank where an anaerobic condition permits the release of phosphorus by poly-P bacteria from the return activated sludge (RAS). The released phosphorus is removed

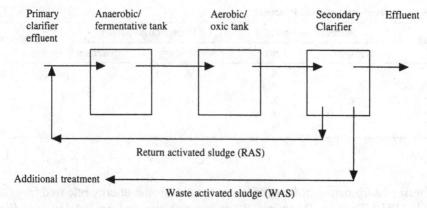

FIGURE 12.7 A/O process.

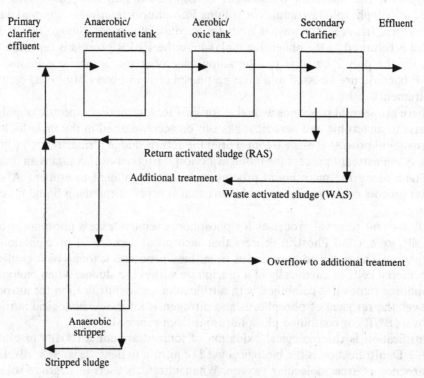

FIGURE 12.8 Phostrip process.

("washed") from the stripper tank by elutriation water. Lime is added to the stripper tank overflow to precipitate the released phosphorus.

There are three combined phosphorus/nitrogen removal processes that are marketed in the United States. These processes included the A²/O, five-stage Bardenpho, and UCT.

The A²/O process is licensed by I. Kruger. The A²/O is the acronym for the three operational conditions or tanks employed in the treatment process (Figure 12.9).

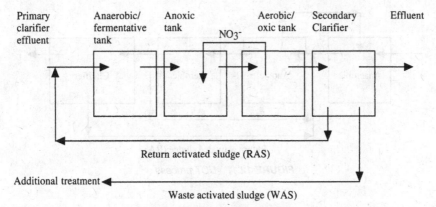

FIGURE 12.9 *A²/O process.*

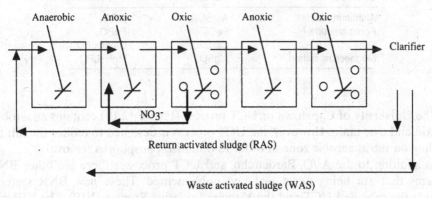

FIGURE 12.10 *Five-stage Bardenpho process.*

These tanks are the anaerobic, anoxic, and oxic. The A²/O process has a relatively low solids retention time (SRT) and high organic loading. As the wastewater and bacteria pass through these tanks, phosphorus is removed biologically and nitrogen is removed through nitrification and denitrification.

In the anaerobic tank fermentation occurs and phosphorus is released to the bulk solution by poly-P bacteria as they absorb and polymerize soluble fatty acids into PHB. In the anoxic tank, facultative anaerobic bacteria use nitrate to degrade soluble cBOD. In the oxic tank, ionized ammonia in the wastewater and released during degradation of organic nitrogen compounds are oxidized to nitrate. In the anoxic tank and oxic tank phosphates are removed from the bulk solution as PHB is solubilized and degraded.

The five-stage or modified Bardenpho process was developed by Barnard in South Africa in 1975 and is licensed and marketed in the United States by Eimco Process Equipment Company. This process is a modification of the original Bardenpho process due to the incorporation of an anaerobic tank upstream of two anoxic/oxic tanks (Figure 12.10). The anoxic tank uses influent wastewater as a carbon source for denitrification.

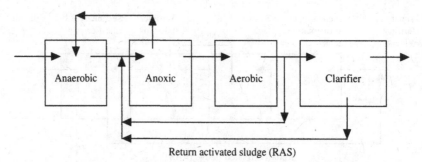

Return activated sludge (RAS)

FIGURE 12.11   UCT process.

TABLE 12.7   Chemicals Commonly Used for Orthophosphate Precipitation and Their Products

Chemical	Formula	Product
Aluminum sulfate	$Al_2(SO_4)_3$	$AlPO_4$
Ferric chloride	$FeCl_3$	$FePO_4$
Lime	$Ca(OH)_2$	$Ca_2OH(PO_4)_3$
Magnesium sulfate	$MgSO_4$	$MgNH_4PO_4$

The University of Capetown or UCT process (Figure 12.11) contains anaerobic, anoxic, and oxic tanks. However, the UCT process is designed to reduce the nitrate loading on the anaerobic zone in order to optimize phosphorus removal.

In addition to the A^2/O, Bardenpho, and UCT processes there are other BNR systems that are being developed and implemented. These new BNR systems include the modified UCT and the Virginia Initiative Process (VIP). The VIP was developed by Hampton Roads Sanitation District and CH2M. The selection of the EBPR or BNR system used at wastewater treatment plants is based upon cost, wastewater composition, and nutrient removal requirements.

PHOSPHORUS REMOVAL BY CHEMICAL ADDITION

Chemical precipitation of orthophosphate is commonly practiced at wastewater treatment plants. Although polyphosphates and organic phosphorus compounds are not removed by chemical precipitation, they are hydrolyzed and mineralized (degraded) to release orthophosphate, which is then chemically precipitated.

Metals that are commonly used to precipitate orthophosphate are Al^{3+}, Ca^{2+}, Fe^{3+}, and Mg^{2+} (Table 12.7). Chemical precipitation of orthophosphate is controlled by pH—for example, pH > 8.5 for Ca^{2+} and pH 5.5 to 7.0 for Al^{3+} and Fe^{3+}. Chemical addition to precipitate orthophosphate may be used upstream or downstream of the aeration tank. However, upstream precipitation of orthophosphate may result in a nutrient deficiency for orthophosphate in the aeration tank.

13

Sulfur-Oxidizing and Sulfur-Reducing Bacteria

Although sulfur (S) accounts for <1% of the dry weight of most organisms, sulfur is an essential element for all organisms. Basic compounds that are necessary for all organisms are proteins, and sulfur is found in most proteins. Sulfur also is needed for the production of enzymes and enzyme cofactors such as thiamin and biotin. The sulfur requirement for organisms can be noted in their carbon-to-sulfur ratio. For most bacteria in wastewater treatment processes, the carbon-to-sulfur ratio is 100:1. For methane-forming bacteria, this requirement is higher.

Bacterial cells contain a large number and variety of proteins that fulfill many critical functions including structural and enzymatic roles. Proteins are made up from about 20 different amino acids. Two of these amino acids, cysteine and methionine (Figure 13.1), contain sulfur and are referred to as the sulfur amino acids. Sulfur in cysteine occurs in a thiol group (-SH). The thiol group is known also as the sulhydral, sulhydryl, or hyrosulfide group. Nearly all proteins contain at least one of the sulfur amino acids.

The linkage of oxidized thiol groups of different cysteine molecules determines the configuration or shape of the protein molecule (Figure 13.2). The shape of the protein molecule is directly relevant to the structural or enzymatic role of the protein molecules. When thiol group reagents or toxicants such as lead (Pb) and mercury (Hg) react with the nonoxidized thiol groups, the activity of many enzymes is destroyed or inhibited.

Sulfur can exist in a number of oxidation states (Table 13.1). All bacteria contain sulfur, and most aerobic bacteria and facultative anaerobic bacteria obtain sulfur from the environment in the most oxidized form, sulfate (SO_4^{2-}). However, most

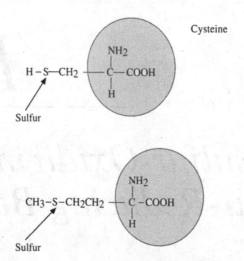

FIGURE 13.1   *Sulfur amino acids, cysteine and methionine. In addition to amino groups (-NH₂) and carboxyl groups (-COOH), the sulfur amino acids contain the thiol group (-SH).*

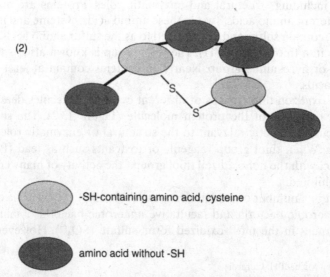

FIGURE 13.2   *Linkage of thiol (-SH) groups. When the thiol groups (-SH) of sulfur amino acids (1) are ionized and joined together, they change the shape of the proteinaceous molecule (2).*

TABLE 13.1 Oxidation States of Sulfur

Oxidation State	Name	Formula
+6	Sulfate	SO_4^{2-}
+5	Dithionate	$S_2O_6^{2-}$
+4	Sulfite	SO_3^{2-}
+4	Disulfite	$S_2O_5^{2-}$
+3	Dithionite	$S_2O_4^{2-}$
+2	Thiosulfate	$S_2O_3^{2-}$
0	Elemental sulfur	S^0
−2	Sulfide	S^{2-}

organic molecules that contain sulfur and are found in bacterial cells contain sulfur in the reduced form as the thiol group (-SH) or disulfide group (-S-S-). These reduced forms are found in cysteine and methionine, respectively. Therefore, bacteria that use sulfate must be able to reduce sulfate.

The sulfur nutrient for most aerobic bacteria and facultative anaerobic bacteria is sulfate (SO_4^{2-}), while the sulfur nutrients for bacteria growing under anaerobic conditions include sulfide (S^{2-}), elemental sulfur (S^0), thiosulfate ($S_2O_3^{2-}$), sulfite (SO_3^{2-}), and the sulfur amino acids.

Sulfate, the most oxidized state of sulfur, is an oxidizing agent and one of the most stable forms of sulfur. Sulfate is the most prevalent anion in natural water and the most abundant form of sulfur in the environment for use by organisms. Sulfite and thiosulfate, lower oxidation states of sulfur, are often found in waterlogged (anaerobic) soils. Sulfite and thiosulfate contribute to malodors associated with these soils.

The sulfur amino acids contain sulfur in the −2 oxidation state. This oxidation state is the same as that found in hydrogen sulfide (H_2S) and sulfides. Therefore, if an organism uses an inorganic sulfur compound with an oxidation state greater than −2 for the synthesis of the sulfur amino acids, the sulfur must be reduced to the −2 oxidation state.

There are two metabolism processes for the reduction of sulfate. These processes are assimilatory sulfate reduction and dissimilatory sulfate reduction (Figures 13.3 and 13.4). During assimilatory sulfate reduction, inorganic sulfate (or any inorganic sulfur with an oxidation state greater than −2) is reduced to sulfide. Sulfide is then used to form the sulfur amino acids (Equation 13.1). Most aerobic bacteria use sulfate as their source for the sulfur nutrient. When sulfate is used, the bacteria release very little sulfide to the environment and do not store sulfide within their cells.

Sulfate + 2 electrons → Sulfite + 6 electrons → Sulfide → Sulfur amino acids   (13.1)

SULFATE-REDUCING BACTERIA

During dissimilatory sulfate reduction relatively large quantities of sulfate are removed from the bulk solution by sulfate-reducing bacteria (SRB). The principal SRB belong in the genera *Desulfovibrio* and *Desulfotomaculum*. Sulfate is removed

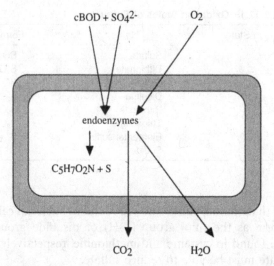

FIGURE 13.3 Assimilatory sulfate reduction. During aerobic conditions, aerobic bacteria remove sulfate from the bulk solution and use sulfate as the sulfur nutrient. Here, sulfate is reduced intracellularly to sulfide (-SH), and the sulfide is incorporated into new cellular material (MLVSS).

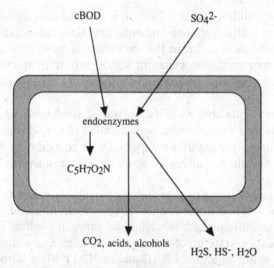

FIGURE 13.4 Dissimilatory sulfate reduction. During dissimilatory sulfate reduction, sulfate-reducing bacteria remove sulfate from the bulk solution in order to degrade soluble cBOD. Sulfate removes from the cell the electrons that are released from the degraded cBOD. Here, the sulfur in the sulfate is not incorporated into new cellular material but is released from the cell in the form of hydrogen sulfide (H_2S) or sulfide (HS^-).

from the bulk solution under strictly anaerobic conditions and is used to oxidize hydrogen (Equation 13.2) and soluble organic compounds such as succinate ($HOOCCH_2CH_2COOH$). SRB produce hydrogen sulfide. Some of the hydrogen sulfide escapes to the atmosphere.

$$SO_4^{2-} + 4H_2 \rightarrow S^{2-} + 4H_2O \tag{13.2}$$

The quantity of hydrogen sulfide that escapes to the atmosphere depends upon the pH of the wastewater, initial dissolved hydrogen sulfide concentration, and temperature. At pH 7, hydrogen sulfide represents 50% of the dissolved sulfides in the wastewater. The concentration of dissolved sulfides as hydrogen sulfide increases as the pH decreases or temperature decreases.

Dissimilatory sulfate reduction also is known as respiratory sulfate reduction. It occurs in waterlogged soils, stagnant ponds, the intestinal tract of ruminant animals, and wherever a strictly anaerobic condition occurs in the presence of sulfate, SRB, and hydrogen or the presence of soluble cBOD. Often, these anaerobic conditions occur in sanitary sewers and wastewater and sludge treatment and holding tanks.

In wastewater treatment plants, dissimilatory sulfate reduction is of concern in anaerobic digesters and treatment units where oxygen and nitrate are absence. SRB are known for their production of malodors and corrosion of iron pipes. SRB reduce sulfate when soluble organic compounds are available and free molecular oxygen and nitrate (NO_3^-) are not available.

When sulfur-containing compounds such as proteins are degraded with the use of sulfate, numerous sulfur-containing, volatile malodorous compounds are produced. Significant malodorous compounds are the mercaptans (Table 13.2). Other sulfur-containing, volatile malodorous compounds include thiocresol (CH_3-C_6H_4-SH), thiophenol (C_6H_5SH), polysulfides (dimethyl disulfide (CH_3-S-S-CH_3), methylethyl disulfide (CH_3-S-S-CH_2CH_3), organic sulfides, and sulfur dioxide (SO_2). Examples of organic sulfides include dially sulfide, dimethyl sulfide, methyl isopropyl sulfide, diethyl sulfide, and methyl pentyl sulfide.

Sulfur dioxide may dissolve in wastewater to produce sulfurous acid (H_2SO_3), hydrogen sulfite (HSO_3^-), and sulfite (SO_3^{2-}). Sulfurous acid dissociates to form hydrogen sulfite, which reacts with free chlorine and combined chorine resulting in the formation of chloride and sulfate. The formation of sulfite produces an increased chlorine demand. Sulfite also is an oxygen scavenger (Equation 13.3) and is found in some industrial wastewaters. Sodium sulfite (Na_2SO_3) often is added to boiler feedwater as a corrosion inhibitor.

$$2SO_3^{2-} + O_2 \rightarrow 2SO4^{2-} \tag{13.3}$$

TABLE 13.2 Mercaptans Commonly Produced During Anaerobic Degradation of Sulfur-Containing Compounds

Mercaptan	Formula	Odor
Allyl	$CH_2{=}CHCH_2SH$	Garlic
Amyl	$CH_3CH_2CH_2CH_2SH$	Putrid
Croty1	$CH_3CH{=}CHCH_2SH$	Skunk
Ethyl	CH_3CH_2SH	Decayed cabbage
Methyl	CH_3SH	Decayed cabbage
t-butyl	$(CH_3)_3CSH$	Skunk

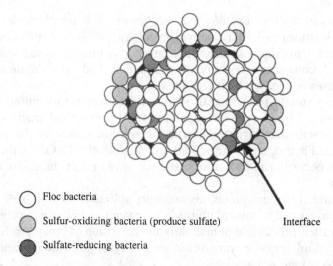

○ Floc bacteria

◐ Sulfur-oxidizing bacteria (produce sulfate)   Interface

● Sulfate-reducing bacteria

FIGURE 13.5 *Interface between oxygen and sulfate; interface between sulfur-oxidizing and sulfate-reducing bacteria. On the interface of a floc particle or biofilm where dissolved oxygen is absent and sulfate is present, numerous sulfur-oxidizing bacteria and sulfate-reducing bacteria may be found. On the outside of the interface where dissolved oxygen is present, sulfur-oxidizing bacteria oxidize reduced forms of sulfur to sulfate. On the inside of the interface where dissolved oxygen is absence and sulfate is present, sulfate-reducing bacteria reduce sulfate to sulfide.*

SULFUR-OXIDIZING BACTERIA

Sulfur-oxidizing bacteria oxidize inorganic sulfur in low oxidation states by adding oxygen to the inorganic sulfur. Sulfur-oxidizing bacteria obtain energy from the oxidation of inorganic sulfur.

Some sulfur-oxidizing bacteria are chemoautotrophs and use the energy that they obtain from the oxidation of inorganic sulfur for the synthesis of organic compounds from carbon dioxide (CO_2). Sulfur-oxidizing chemoautotrophs include species in the genera *Thiobacillus*, *Thiospirillopsis*, and *Thiovulum*. Because most reduced forms of sulfur required by these bacteria are produced by the metabolic activity of SRB, the sulfur-oxidizing chemoautotrophs usually grow to large numbers at the interface between anaerobic conditions (source of sulfide) and aerobic conditions (source of oxygen) (Figure 13.5).

Sulfur-oxidizing bacteria like SRB also are known for operational problems associated with their growth. They convert reduced forms of sulfur to sulfuric acid (H_2SO_4). Concrete (a stone aggregate bonded by calcium carbonate ($CaCO_3$) corrodes or "rots" in the presence of sulfuric acid. Even with the development of sulfuric acid, sulfur-oxidizing bacteria continue to growth. For example, *Thiobacillus* is very tolerant of sulfuric acid and can grow at pH values less than 1.

FILAMENTOUS SULFUR BACTERIA

There are three filamentous sulfur bacteria that are found in activated sludge processes (Table 13.3). These filamentous organisms are *Beggiatoa*, *Thiothrix*, and

TABLE 13.3 Filamentous Sulfur Bacteria and Operational Conditions Associated with Their Rapid Growth

Filamentous Organism	Operational Condition Associated with Rapid Growth
Beggiatoa	Septicity/sulfides
Thiothrix	Low F/M, organic acids, readily degradable cBOD, septicity/sulfides
Type 021N	Low F/M, organic acids, readily degradable cBOD, septicity/sulfides

type 021N. *Beggiatoa* is a white or colorless filamentous organism and is associated more commonly with fixed film processes such as trickling filters than suspended growth systems (activated sludge processes). In trickling filter systems, these organisms usually live at the interface between an aerobic zone and an anaerobic zone.

Rapid and undesired growth of *Thiothrix* and type 021N often occurs in activated sludge processes and contributes to settleability problems and loss of solids from secondary clarifiers. Contributing factors for sulfide production include organic overloading, low dissolved oxygen concentration, and high influent sulfate concentration. Techniques that can be used to control the growth of these filamentous organisms include reducing sulfide concentration to <5 mg/liter and reducing organic loading.

A high sulfide concentration in the aeration tank is a major operational condition. Sulfide is an oxygen scavenger, and its presence results in a low dissolved oxygen concentration in the aeration tank. Sulfide also is responsible for the rapid and undesired growth of filamentous sulfur bacteria that use sulfide as an energy source. They remove sulfide from the bulk solution and oxidize it to elemental sulfur in order to obtain energy.

Elemental sulfur is stored as intracellular granules by these organisms. The granules are surrounded by an envelope, and under phase contrast microscopy the granules appear as highly refractive, bright yellow bodies.

PURPLE SULFUR BACTERIA

The growth of purple sulfur bacteria is encouraged by the presence of hydrogen sulfide. Purple sulfur bacteria are photosynthetic and include the genera *Chromatium*, *Thiocapsa*, and *Thiopedia*. These bacteria oxidize sulfides to elemental sulfur to obtain energy and deposit the sulfur intracellularly as granules. In dense populations the purple sulfur bacteria can turn wastewater red.

Oxygen is toxic to purple sulfur bacteria. Therefore, the growth of these bacteria is arrested or prevented in the presence of free molecular oxygen.

SOURCES OF SULFUR

Sulfur is found in many different compounds of natural and pollutant origin and is very common in wastewater. Domestic wastewater contains approximately 3–6 mg/liter of organic sulfur as proteinaceous wastes and approximately 4 mg/liter of organic sulfur as sulfonates derived from detergents. Domestic wastewater also con-

TABLE 13.4 Important Sulfur-Containing Functional Groups

Name	Structure
Disulfide	–S–S–
Sulfonic acid	–SOOOH
Sulfoxide	–S– $\parallel$ O
Thioketone	–C– $\parallel$ S
Thiol/sulfide	–SH

tains approximately 20 mg/liter of sulfate. In sludge, total sulfur may be 1–2% of the dry weight of the sludge.

Sulfur is excreted by aerobic organisms as sulfate. However, upon the death of these organisms, sulfur is released from the organisms in the reduced state. Reduced sulfur or sulfide is produced from the oxidized forms of sulfur (sulfate, sulfite, thiosulfate, and elemental sulfur) during anaerobic degradation of substrate by anaerobic bacteria such as *Desulfovibrio desulphuricans*.

The degradation of sulfur-containing compounds often has an adverse impact upon wastewater treatment plants and water quality. Biological compounds that contain sulfur include amino acids and proteins. Sulfur is present in natural waters and wastewater in many functional groups (Table 13.4).

Because a large variety of organic sulfur compounds exist, numerous sulfur products and biochemical reactions are associated with the degradation of organic sulfur compounds. The degradation of sulfur-containing amino acids can result in the production of volatile sulfur compounds (VSC). In addition to hydrogen sulfide, major VSC associated with the degradation of sulfur-containing organic compounds are methyl thiol (CH_3SH) and dimethyl disulfide (CH_3SSCH_3).

Industrial wastewaters also may contain large quantities of sulfur-containing compounds. Industrial wastewaters of concern are those that contain sulfonated detergents and sulfite waste liquor.

SULFONATED DETERGENTS

A large variety of detergents including alkyl benzene sulfonates (ABS) are used for commercial, domestic, and industrial purposes. When present in wastewater, ABS represent a source of sulfur and like many other detergents, ABS also represent several additional operational concerns including:

- Decreased rate of oxygen dissolution
- Proliferation of anaerobic bacteria
- Resistance to biological degradation
- Toxicity

WOOD-PULPING

The manufacturing of wood pulp for the production of paper involves the digestion of wood with steam, high pressure, and calcium bisulfite ($Ca(HSO_3)_2$) or magnesium bisulfite ($Mg(HSO_3)_2$). The digestion of wood removes most of the lignin (a complex phenolic polymer) and other compounds from the wood but does not remove the cellulose or pulp. The lignin and other removed compounds are sulfonated by the bisulfite and may be released into solution. The waste material containing sulfonated lignin, residual bisulfite, and other compounds is known as sulfite waste liquor.

SULFUR BACTERIA

There are five major groups of sulfur bacteria (Table 13.5). These groups include the sulfate-reducing bacteria (SRB), sulfur-oxidizing bacteria, colorless sulfur bacteria, sulfur-oxidizing photosynthetic green bacteria, and sulfur-oxidizing photosynthetic purple bacteria. Growth media and identification of sulfur bacteria are provided in the latest edition of *Standard Methods for the Examination of Water and Wastewater*.

Sulfur bacteria, especially SRB, often grow with other bacteria, and their presence may not easily be detected by microscopic examination. SRB usually are outnumbered by other groups of bacteria, except in special environments such as anaerobic digesters. However, changes in the sulfur profile including sulfate and sulfides across a treatment unit may be indicative of the presence of specific sulfur bacteria.

Operational concerns related to the activity of SRB include malodor production, undesired growth of the filamentous organisms *Beggiatoa*, *Thiothrix*, and type 021N, and competition with methane-producing bacteria.

If sulfates are present in anaerobic digester sludge, *Desulfovibrio desulfuricnas* proliferates. This SRB reduces sulfate to hydrogen sulfide by using hydrogen. Methane-forming bacteria also use hydrogen to produce methane. The presence of sulfate in anaerobic digester sludge results in competition for hydrogen by SRB and methane-producing bacteria. The methane-producing bacteria accordingly use less substrate and produce less methane. In addition, the hydrogen sulfide produced by the SRB has an inhibitory effect on methane-producing bacteria.

When SRB use sulfate to degrade substrate such as ethanol, they produce acetate. Here, SRB act as a competitor for ethanol with methane-producing bacteria and substrate producer for methane-producing bacteria.

In the absence of sulfate, many SRB are able to adapt and continue to proliferate. Some grow with hydrogen-consuming, methane-producing bacteria. When this occurs, the SRB produce substrates such as acetate and hydrogen that can be used by methane-producing bacteria.

In the absence of free molecular oxygen or nitrate or presence of an oxygen and nitrate gradient, SRB use low-molecular-weight carbon sources as substrates. These carbon sources are produced by fermentation of carbohydrates, lipids, and proteins (Table 13.6).

Sulfides discharged to an anaerobic digester or produced in an anaerobic digester combined with soluble metals such as cadmium, iron, and zinc to form highly insol-

TABLE 13.5   Major Groups of Sulfur Bacteria

Sulfate-reducing bacteria (SRB)
  Desulfobacter
  Desulfobacterium
  Desulfococcus
  Desulfomonas
  Desulfonema
  Desulfosarcina
  Desulfotomaculum
  Desulfovibrio
  Desulfuromonas
  Thermodesulfobacterium
Sulfur-oxidizing bacteria
  Arthrobacter
  Bacillus
  Micrococcus
  Pseudomonas
  Thiobacillus
  Thiospirillopsis
  Thiovulum
Colorless sulfur bacteria
  Beggiatoa
  Thiothrix
  Type 021N
Sulfur-oxidizing photosynthetic green bacteria
  Anacalochloris
  Chlorobium
  Pelodictyon
Sulfur-oxidizing photosynthetic purple bacteria
  Amoebobacter
  Chromatium
  Ectothiorhodospira
  Lamprobacter
  Lamprocystis
  Thiocapsa
  Thiocystis
  Thiodictyon
  Thiopedia
  Thiospirillum

TABLE 13.6   Carbon Sources Used as Substrate by SBR

Name	Formula
Acetate	CH_3COOH
Ethanol	CH_3CH_2OH
Fumarate	$HOOCCHCOOH$
Lactate	$CH_3CH(OH)COOH$
Propanol	$CH_3CH_2CH_3OH$
Propionate	CH_3CH_2COOH
Pyruvate	$CH_3COCOOH$
Succinate	$HOOCCH_2CH_2COOH$

uble salts. Additional metals that combined with soluble sulfide include chromium, copper, lead, and nickel. Examples of sulfide salts include cadmium sulfide (CdS), ferrous sulfide (FeS), and zinc sulfide (ZnS). These metallic sulfides and immobilized heavy metals turn sludge black. Hydrogen sulfide does not tend to accumulate in an anaerobic digester, until the metals are removed from solution.

Hydrogen sulfide may be chemically and biologically oxidized to sulfate. Hydrogen sulfide is an oxygen scavenger; when exposed to air or oxygenated water, it may be spontaneously converted to sulfate. Hydrogen sulfide may be biologically oxidized to sulfate by sulfur-oxidizing bacteria.

There are two commonly occurring sulfur-oxidizing bacteria that are important in malodor control of wastewater. The bacteria are *Chlorobium* and *Chromatium*. These photosynthetic bacteria oxidize sulfides to sulfur compounds that are not malodorous. When present in highly concentrated masses, *Chromatium* gives a red color to wastewater in the presence of organic overloading and septicity.

In the activated sludge process the sulfur filamentous organisms—type 021N, *Beggiatoa*, and *Thiothrix*—oxidize sulfides to elemental sulfur (Equation 13.4), and *Thiobacillus* oxidizes elemental sulfur to sulfate (Equation 13.5).

$$S^{2-} + \tfrac{1}{2}O_2 + 2H^+ \rightarrow S^0 + H_2O \tag{13.4}$$

$$2S^0 + 3O_2 + 2H_2O \rightarrow H_2SO_4 \tag{13.5}$$

MOVEMENT OF SULFUR THROUGH AN ACTIVATED SLUDGE PROCESS

Sulfur enters a wastewater treatment plant in inorganic and organic forms (Figure 13.6). Sulfides as H_2S and HS^- are present in the sewer system as the result of two biochemical events. First, thiol groups (-SH) are released from the sulfur amino acids and sulfur-containing proteins (organic sulfur compounds) when these compounds are degraded by bacteria in the biofilm and sediment of the sewer system. Second, H_2S and HS– are produced in the sewer system when sulfate (SO_4^{2-}) is reduced through dissimilatory sulfate reduction by SRB when they degrade soluble cBOD.

The anaerobic degradation of organic sulfur compounds in the sewer system also results in the production of several malodorous, sulfur-containing organic compounds. Some of these compounds may be released to the atmosphere.

Sulfate is present in raw wastewater. The significant sources of sulfate are urine and groundwater. Sulfite (SO_3^{2-}) may be present in raw wastewater, if it is discharged from boiler feedwater or cooling tower water.

The primary clarifier influent contains inorganic and organic forms of sulfur. Inorganic forms consist of H_2S, HS^-, SO_4^{2-}, and perhaps SO_3^{2-}, while organic forms consist of the sulfur amino acids, sulfur-containing proteins, and malodorous sulfur-containing compounds. Some of the influent proteins may be removed in the primary clarifier, if they are adsorbed to solids that settle out. Some hydrogen sulfide and some malodorous sulfur-containing compounds may be released from the primary clarifier to the atmosphere.

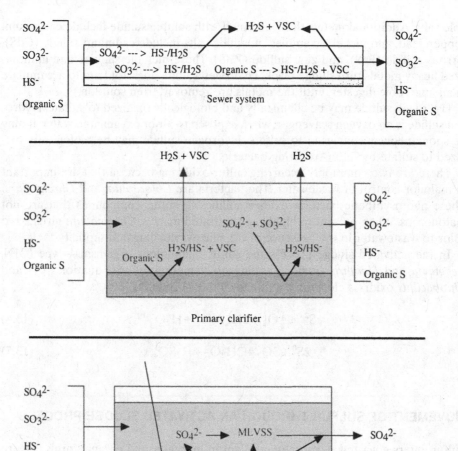

FIGURE 13.6 *Movement of sulfur in the activated sludge process. Sulfur compounds enter the sewer system as sulfate in groundwater and urine, sulfide and sulfite as oxygen scavengers, and organic sulfur. In the sewer system under anaerobic conditions, sulfate and sulfite are reduced to sulfide through the degradation of cBOD. Some sulfide escapes the sewer system as hydrogen sulfide, while sulfur-containing malodorous compounds from the degradation of amino acids and proteins also escape the sewer system. In the primary clarifier, additional hydrogen sulfide and sulfur-containing malodorous compounds may be released; with increasing HRT in the primary clarifier, sulfate may be used to degrade cBOD in the sludge, and fermentation may produced sulfur-containing malodorous compounds. In the aeration tank, sulfide and sulfite may be biologically and chemically oxidized to sulfate. Sulfide also may be oxidized to elemental sulfur by sulfur-oxidizing filamentous organisms such as type 021N. VSC produced in the sewer system, primary clarifier, and secondary clarifier may be stripped to the atmosphere; with increasing HRT in the aeration tank, organic sulfur compounds are degraded to release sulfide that are oxidized to sulfate.*

Most sulfur compounds that enter the primary clarifier can be found in the primary clarifier effluent, because they are either soluble or suspended. These compounds then enter the aeration tank of the activated sludge process.

In the aeration tank, numerous biological and chemical changes occur with the sulfur compounds. Sulfides in the aeration tank may be biologically and chemically oxidized to sulfate. These oxidations result in dissolved oxygen consumption. Some sulfides may be removed by the sulfur filamentous organisms (type 021N, *Beggiatoa*, and *Thiothrix*) and oxidized to elemental sulfur (S^0)

Malodorous sulfur-containing compounds may be stripped to the atmosphere or degraded. Their degradation results in the release of sulfides. With sufficient hydraulic retention time (HRT) the sulfur amino acids that are absorbed and degraded and the sulfur-containing proteins that are adsorbed, solubilized, and degraded release thiol groups. The released sulfides and thiol groups also undergo biological and chemically oxidation to sulfate.

Bacteria in the aeration tank remove sulfate as their sulfur nutrient. Here, the sulfate is reduced intracellularly to sulfide and incorporated into cellular material (MLVSS) or organic sulfur.

Activated sludge solids and nondegraded cBOD that are transferred to a secondary clarifier or thickener and remain for a relatively long retention time can experience anaerobic activity, if free molecular oxygen and nitrate (NO_3^-) are not available for bacterial activity. Anaerobic activity results in the use of sulfate and fermentative pathways to degrade cBOD. As a consequence of anaerobic activity, H_2S, HS^-, and malodorous sulfur-containing compounds are produced. Soluble forms of sulfur leave the secondary clarifier, while soluble, colloidal, and particulate forms of sulfur are returned from the thickener overflow to the activated sludge process.

MOVEMENT OF SULFUR THROUGH AN ANAEROBIC DIGESTER

Sulfur compounds that are transferred to an anaerobic digester also undergo numerous biological and chemical events (Figure 13.7). Sulfides that enter the digester or released in the digester through anaerobic degradation of sulfur-containing compounds or dissimilatory sulfate reduction may be

- Removed from solution by anaerobic bacteria as their sulfur nutrient,
- Bonded and precipitated from solution by soluble metals such as cadmium, iron, and zinc,
- Sequestered (chelated) and held in solution by fatty acids that are produced in the digester through fermentative processes, or
- Released to the biogas as H_2S from the sludge.

Organic sulfur compounds (amino acids and proteins) are degraded through anaerobic activity. Their degradation results in the release of thiol groups that experience the same fates in the digester as sulfide.

Sulfate in the anaerobic digester is reduced to sulfide through dissimilatory sulfate reduction. However, an excessive quantity of sulfate represents three poten-

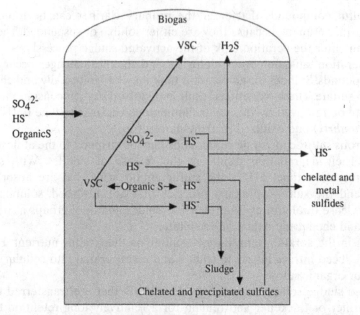

FIGURE 13.7   Movement of sulfur in the anaerobic digester. Sulfur compounds enters the anaerobic digester as sulfate, sulfide, and organic sulfur. The anaerobic degradation of organic sulfur compounds results in the release of sulfide and VSC. Some of the VSC are degraded, while some VSC are collected in the biogas. Sulfide can be used as the sulfur nutrient for bacteria during anaerobic activity resulting in increased sludge production and may be chelated to organic acids or precipitated in metal salts. Sulfate is reduced to sulfide by sulfate-reducing bacteria.

TABLE 13.7   Guideline ORP and Anaerobic Digester Activity

ORP (mV)	Anaerobic Digester Activity
<−100	Sulfate reduction of substrate
<−200	Fermentation, mixed acid and mixed alcohol production
<−300	Methane production

tial operational concerns with respect to digester performance. These concerns are (1) an increase in oxidation-reduction potential, (2) production of toxic hydrogen sulfide, and (3) a decrease in available substrate for methane-producing bacteria.

The presence of sulfates in an anaerobic digester increases the oxidation–reduction potential (ORP) of the digester sludge (Table 13.7). With increasing ORP, methane-producing bacteria become less active and fatty acids, one of primary substrates for methane-forming bacteria, accumulate. Although methane-forming bacteria become less active with increasing ORP, fermentative bacteria and SRB do not. These two bacterial groups continue to produce fatty acids. The accumulating fatty acids destroy alkalinity and lower pH, and the digester becomes "sour."

Although H_2S/HS^- at a low concentration is beneficial to an anaerobic digester, because it serves as the sulfur nutrient for anaerobic bacteria including methane-forming bacteria, at a high concentration it is toxic. With an increasing quantity of sulfides from anaerobic activity and decreasing pH (<7), toxic H_2S is formed.

SRB and methane-forming bacteria have similar characteristics. Both are active under strict anaerobic conditions with similar pH and temperature growth ranges. Like methane-forming bacteria, some SRB are able to oxidize hydrogen (H_2) and acetate (CH_3COOH). Therefore, SRB compete with methane-forming bacteria for these substrates. Competition results in a loss of substrate for methane-forming bacteria and a decrease in methane production.

Part IV

Floc Formation

14

Floc-Forming Bacteria

Floc formation does not occur in an anaerobic digester. Septicity in the digester destroys floc formation. Floc formation does occur in the activated sludge process and is essential for its success. Floc formation permits the "packaging" of a large and diverse population of bacteria in numerous floc particles that (1) can be separated from the waste stream in the secondary clarifier and (2) can be recycled (Figure 14.1) as needed to achieve the following treatment objectives:

- Degrade carbonaceous BOD.
- Degrade nitrogenous BOD.
- Remove fine solids (colloids, dispersed cells, and particulate material).
- Remove phosphorus.
- Develop a diverse population of higher life forms (ciliated protozoa, rotifers, and free-living nematodes) that improve treatment efficiency.

Floc formation occurs naturally with increasing MCRT and is initiated by floc-forming bacteria (Table 14.1). These bacteria are able to produce three necessary cellular components that enable them to "stick" together or agglutinate. These cellular components are (1) pili or fibrils, (2) sticky polysaccharides, and (3) poly-β-hydroxybutyrate (PHB) or starch granules (Figure 14.2).

The pili or fibrils are extensions of the cell membrane that protrude through the cell wall into the bulk solution. The fibrils contain key functional groups such as the carboxyl group (-COOH) and the hydroxyl group (-OH) that become ionized with the lost of the hydrogen atoms. The ionized fibrils or bacterial cells are joined together by bivalent cations such as calcium (Ca^{2+}) that are in solution (Figure 14.3). The joining of fibrils from different bacterial cells initiates floc formation. Ionized

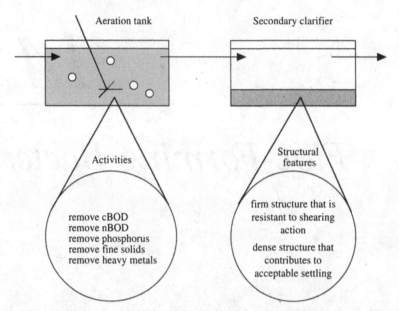

FIGURE 14.1 *Essential activities and structural features of the floc particle in activated sludge. Essential activities performed by the floc particle in the aeration tank include (1) removal of cBOD, (2) removal of nBOD, (3) removal of phosphorus, (4) removal of fine solids, and (5) removal of heavy metals. Necessary structural features of the floc particle that are important in the aeration tank and secondary clarifier include (1) firm structure that is resistant to shearing action or turbulence and (2) dense structure that contributes to acceptable settling.*

TABLE 14.1   Significant Genera of Floc-Forming Bacteria

Achromobacter	Citromonas
Aerobacter	Escherichia
Alcaligenes	Flavobacterium
Arthrobacter	Pseudomonas
Bacillus	Zoogloea

fibrils that are not joined together remain exposed to the bulk solution and act as the "wisps of a broom" as they sweep and remove fine solids and heavy metals from the bulk solution.

There are several polysaccharides that make up the glycocalyx or sticky coating outside the bacterial cell wall and contribute to floc formation by "sticking" cells together (Figure 14.4). Some polysaccharides, such as those produced by young bacterial cells, are weak-bonding and produced in large quantities. Other polysaccharides, such as those produced by old bacterial cells, are strong-bonding and produced in small quantities. The differences in bonding strength and quantity of polysaccharides produced result in the development of (a) weak and buoyant floc particles at a young sludge age and (b) firm and dense floc particles at an old sludge age.

Poly-β-hydroxybutyrate (PHB) is a starch and serves two purposes. First, it is stored inside and outside the bacterial cell where it serves as a food reserve (Figure 14.5). Second, when stored outside the cell, PHB helps to anchor bacterial cells more

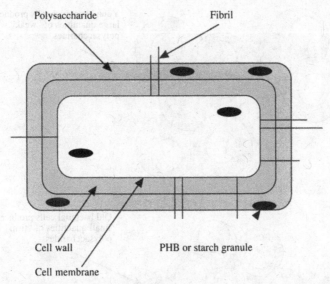

FIGURE 14.2 *Necessary cellular components for the initiation of floc formation. There are three necessary cellular components for floc formation. These components include bacterial fibrils, "sticky" polysaccharides, and PHB or starch granules.*

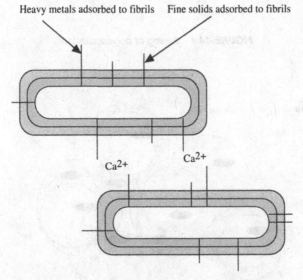

FIGURE 14.3 *Joining of bacterial fibrils. The ionized fibrils on bacterial cells are joined together by bivalent cations such as calcium (Ca^{2+}) that are in solution.*

tightly together; that is, it improves floc formation. When present at the perimeter of the floc particle, PHB also helps to anchor particulate material to floc particles.

PHB production and deposition outside bacterial cells is slow. Therefore, rapidly growing, young floc particles have PHB mostly in the core of the floc particle, while slowly growing, old floc particles have PHB in large quantities in the core and

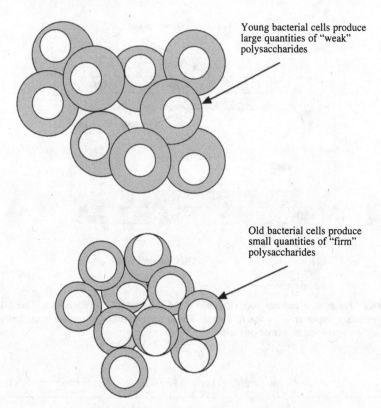

Young bacterial cells produce large quantities of "weak" polysaccharides

Old bacterial cells produce small quantities of "firm" polysaccharides

FIGURE 14.4 *Joining of polysaccharides.*

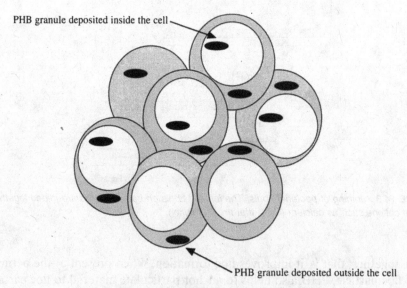

PHB granule deposited inside the cell

PHB granule deposited outside the cell

FIGURE 14.5 *Polyhydroxybutyrate (PHB) deposition.*

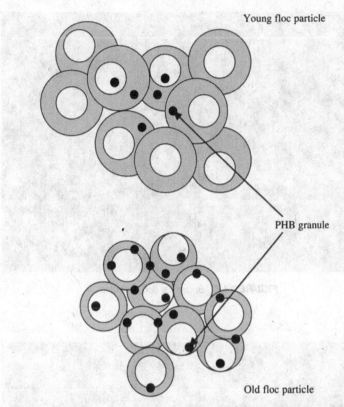

FIGURE 14.6 *Deposition of PHB granules in young and old floc particles.*

perimeter of the floc particles (Figure 14.6). Due to the differences in quantity and location of PHB, old floc particles or sludge is stronger and denser than young floc particles or sludge.

Filamentous organisms also perform a significant role in floc formation. They provide an internal "backbone" or network of strength for the floc particle that enables the particle to resist shearing action and increase in size. The increase in size provides for a larger number and diversity of bacteria and improved treatment efficiency.

The increase in size of the floc particles occurs as bacteria grow along the length of the filamentous organisms. Because floc formation is initiated before filamentous organisms can increase in length and extend beyond the perimeter of the floc particle, young floc particles have little or no filamentous organisms as compared to old floc particles. Due to the absence of filamentous organisms, young floc particles are spherical in shape (Figure 14.7), while old floc particles that contain filamentous organisms are irregular in shape (Figure 14.8).

There are significant differences between young floc particles or sludge and old floc particles or sludge (Table 14.2). These differences are responsible for the production of weak and buoyant floc particles at a young sludge age and firm and dense floc particles at an old sludge age. The size of a floc particle is determined by (1) the agglutinating strength of the floc-forming bacteria, (2) absence or presence of

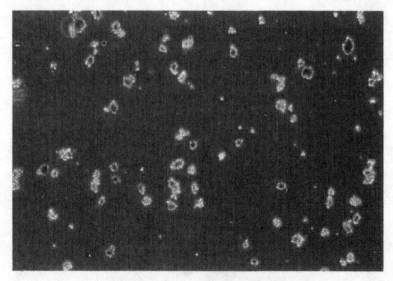

FIGURE 14.7   Spherical floc particles.

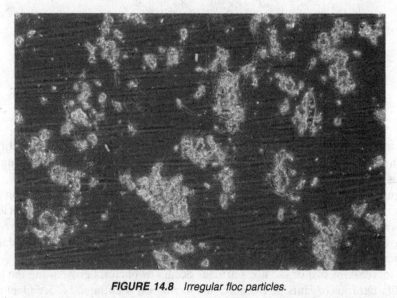

FIGURE 14.8   Irregular floc particles.

TABLE 14.2   Significant Differences between Young Floc Particles and Old Floc Particles

Feature	Young Floc Particles	Old Floc Particles
Shape	Spherical	Irregular
Size	Small (<150 µm)	Medium (150–500 µm) and Large (>500 µm)
Color	Light	Golden-brown
Filamentous organisms	Insignificant	Significant
Strength	Weak	Firm
Settleability	Poor	Good

TABLE 14.3 Operational Conditions Associated with the Interruption of Floc Formation

Operational Condition	Description or Example
Cell bursting agent	Lauryl sulfate
Colloidal floc	Nondegrading or slowly degrading colloids
Elevated temperature	>32°C
Foam production and accumulation	Foam-producing filamentous organisms
High pH	>8.5
Increase in percent MLVSS	Accumulation of fats, oils, and grease
Lack of ciliated protozoa	<100 per milliliter
Low dissolved oxygen concentration	<1 mg/liter for 10^+ consecutive hours
Low pH	<6.5
Nutrient deficiency	Usually nitrogen or phosphorus
Salinity	Excess manganese, potassium, or sodium
Scum production and accumulation	Toxicity and die-off of bacteria
Septicity	ORP < −100 mV
Shearing action	Surface aeration
Slug discharge of soluble cBOD	3X normal quantity of soluble cBOD
Sulfates	>500 mg/liter
Surfactants	Excess anionic detergents
Total dissolved solids	>5000 mg/liter
Toxicity	RAS chlorination
Undesired filamentous organism growth	>5 filaments per floc particle
Viscous floc or Zoogloeal growth	Rapid floc-forming bacterial growth
Young sludge age	<3 days MCRT

filamentous organisms, and (3) turbulence level in the treatment process. The change in color of the floc particles from white to golden-brown is due in large part to the accumulation of secreted oils by bacteria as they age.

Floc formation can be interrupted by several operational conditions (Table 14.3). Interruption of floc formation results in the loss of settleability, the loss of solids, and an increase in operational costs and, possibly, permit violations.

15

Filamentous Bacteria

Although there are some filamentous algae and filamentous fungi in the activated sludge process, most filamentous organisms are bacteria (Table 15.1). Filamentous bacteria enter the activated sludge process through (1) inflow and infiltration as soil and water organisms, (2) their growth in biofilm in sanitary sewers, and (3) the effluent of industrial wastewater treatment plants that biologically pretreat their wastewater.

Filamentous bacteria perform positive roles and negative roles in the activated sludge process. Positive roles include (1) degradation of soluble cBOD, (2) improvement in floc formation, and (3) degradation of some complex forms of cBOD. Negative roles include (1) settleability problems, (2) loss of solids, and (3) foam production. Foam-producing filamentous organisms include *Microthrix parvicella*, Nocardioforms, type 0092, and type 1863.

Whether filamentous bacteria perform positive or negative roles in the activated sludge process is determined by their relative abundance. Positive roles are experienced when one to five filamentous bacteria are present in most floc particles. This level of filamentous growth is equal to a relative abundance rating of "3" or "common" on the table of Relative Abundance Ratings for Filamentous Organisms (Table 15.2). At relative abundance ratings of "0," "1," and "2," filamentous bacteria are not present in adequate numbers to provide significant positive roles. Relative abundance ratings < "3" are due to (1) young sludge age, (2) complex cBOD as the major substrates, and (3) toxicity.

At relative abundance ratings > "3" the negative roles of filamentous bacteria are experienced. The undesired growth of filamentous bacteria at ratings > "3" can be associated with specific operational conditions for each filamentous bacteria (Table 15.3). It may be necessary to perform a Gram stain on a smear of mixed liquor to

TABLE 15.1 Major Filamentous Bacteria in the Activated Sludge Process

Beggiatoa	Type 0581
Haliscomenobacter hydrossis	Type 0675
Microthrix parvicella	Type 0803
Nocardioforms	Type 0961
Nosticoda limicola	Type 1701
Sphaerotilus natans	Type 1702
Thiothrix	Type 1851
Type 0041	Type 1863
Type 0092	Type 021N

TABLE 15.2 Relative Abundance Ratings for Filamentous Organisms

Rating	Term	Description	Undesired Settleability Experience
"0"	"None"	Filaments not observed	No
"1"	"Few"	Filaments present; in an occasional floc	No
"2"	"Some"	Filaments present; frequently observed, but only in some floc	No
"3"	"Common"	Filaments present; 1–5 filaments in most floc	No, unless significant interfloc bridging or open floc formation or foam-producing filaments in floc
"4"	"Very common"	Filaments present; 6–20 filaments in most floc	Yes
"5"	"Abundant"	Filaments present; >20 filaments in most floc	Yes
"6"	"Excessive"	Filaments present; more filaments than floc or filaments are excessive in bulk solution	Yes

determine the relative abundance of the filamentous bacteria, if the filamentous bacteria are (1) translucent, (2) short in length, and (3) mostly within the floc particles. Of the settleability problems and loss of solids issues related to undesired growth of filamentous bacteria, most are due to 10 filamentous bacteria. These bacteria include *Haliscomenobacter hydrossis*, *Microthrix parvicella*, Nocardioforms, *Sphaerotilus natans*, *Thiothrix* spp., type 0041, type 0092, type 0675, type 1701, and type 021N.

In addition to undesired filamentous bacterial growth, there are two filamentous bacteria/floc particle structures that contribute to settleability problems and loss of solids. These forms are interfloc bridging (Figure 15.1) and open floc formation (Figure 15.2). Interfloc bridging is the joining in the bulk solution of the extended filamentous bacteria from the perimeter of two or more floc particles. Open floc formation is the scattering of the floc bacteria in small groups along the lengths of the filamentous bacteria in the floc particle.

There are four foam-producing filamentous bacteria. These organisms are *Microthrix parvicella*, Nocardioforms, type 0092, and type 1863. These organisms

TABLE 15.3 Operational Conditions Associated with the Undesired Growth of Filamentous Bacteria

Operational Condition	Filamentous Bacteria
High MCRT(>10 days)	0041, 0092, 0581, 0675, 0803, 0961, 1851, *Microthrix parvicella*
Fats, oils, and grease	0092, *Microthrix parvicella*, Nocardioforms
High F/M or slug discharge of soluble cBOD	1863
High pH (>8.0)	*Microthrix parvicella*
Low dissolved oxygen and high MCRT	*Microthrix parvicella*
Low dissolved oxygen and low to moderate MCRT	*Haliscomenobacter hydrossis*, *Sphaerotilus natans*, 1701
Low F/M (<0.05)	*Haliscomenobacter hydrossis*, *Microthrix parvicella*, Nocardioforms, 0041, 0092, 0581, 0675, 0803, 0961, 021N
Low nitrogen or phosphorus	*Haliscomenobacter hydrossis*, Nocardioforms, *Sphaerotilus natans*, *Thiothrix*, 0041, 0092, 0675, 1701, 021N
Low pH (<6.5)	Nocardioforms
Organic acids	*Beggiatoa*, *Thiothrix*, 021N
Readily degradable substrates, e.g., alcohols, amino acids with sulfur, glucose, volatile fatty acids	*Haliscomenobacter hydrossis*, *Nosticoda limicola*, *Sphaerotilus natans*, *Thiothrix*, 1851, 021N
Septicity/sulfides (1–15 mg/liter)	*Beggiatoa*, *Nosticoda limicola*, *Thiothrix*, 0041, 021N
Slowly degradable substrates	*Microthrix parvicella*, Nocardioforms, 0041, 0092
Warm wastewater temperature	*Sphaerotilus natans*, 1701
Winter proliferation	*Microthrix parvicella*

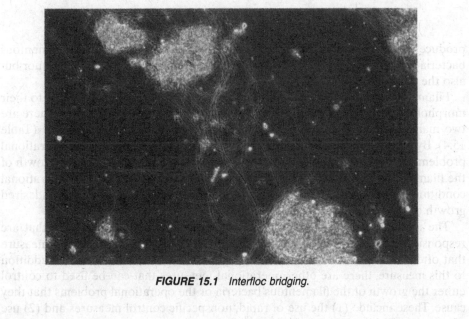

FIGURE 15.1 Interfloc bridging.

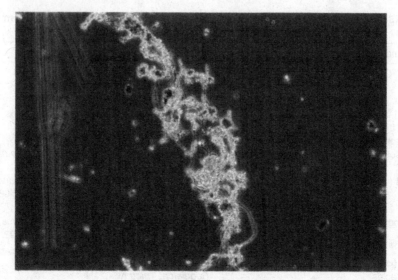

FIGURE 15.2 *Open floc formation.*

TABLE 15.4 Manuals Available for the Identification of Filamentous Organisms

D. H. Eikelboom and H. J. J. van Buijsen. 1989. *Microscopic Sludge Investigation Manual.* TNO
  Research Institute for Environmental Hygiene, Water and Soil Division, P.O. Box 214, 2600 AE
  Delft, The Netherlands.
D. Jenkins, M. Richards, and G. T. Daigger. 2004. *Manual on the Causes and Control of Activated
  Sludge Bulking and Foaming,* 3rd edition. CRC Lewis Publications, Boca Raton, FL.

produce viscous chocolate-brown foam. Control of the growth of these filamentous
bacteria requires appropriate treatment measures for not only the mixed liquor but
also the foam.

Filamentous organisms are identified to name or type number according to their
morphology, response to staining techniques, and ability to oxidize sulfur. There are
two manuals that are used for the identification of filamentous organisms (Table
15.4). By identifying the filamentous organisms that are responsible for operational
problems, the operational conditions that permit the rapid and undesired growth of
the filamentous organisms also can be identified (Table 15.3). Once the operational
conditions have been identified, they may be regulated to control the undesired
growth of the filamentous organisms.

The identification of filamentous bacteria and operational conditions that are
responsible for their undesired growth is part of a slow, specific operational measure
that often is used to control undesired growth of filamentous bacteria. In addition
to this measure, there are other operational measures that can be used to control
either the growth of the filamentous bacteria or the operational problems that they
cause. These include (1) the use of rapid, nonspecific control measures and (2) use
of selectors (Table 15.5).

TABLE 15.5   Operational Measures Available for the Control of Undesired Filamentous Bacteria or Their Operational Problems

Rapid, nonspecific control measures
  Increase the return activated sludge (RAS) rate
  Manipulate the substrate feed point to the aeration tank
  Add a coagulant to the secondary clarifier influent
  Add a polymer to the secondary clarifier influent
  Add a toxicant
Slow, specific control measures
  Identify the problematic filamentous organisms
  Identify the operational conditions responsible for undesired filamentous growth
  Regulate the operational conditions to prevent undesired filamentous growth
Use of selectors
  Anoxic
  Anaerobic
  F/M

RAPID, NONSPECIFIC CONTROL MEASURES

Rapid, nonspecific control measures usually are used without the identification of the problematic filamentous bacteria or their operational conditions that permit their undesired growth. These measures consist of (1) adjustment of RAS rate, (2) manipulation of the substrate feed point to the aeration tank, (3) addition of a coagulant to the secondary clarifier influent, (4) addition of a polymer to the secondary clarifier influent, and (5) the addition of a toxicant.

The first four rapid, nonspecific control measures do not control the growth of the filamentous bacteria but, instead, control the operational problems associated with the undesired growth—that is, settleability problems and loss of solids. The RAS rate may be increased to remove solids more quickly from the secondary clarifier. However, an increase in RAS rate may shear floc particles and decreases the hydraulic retention time (HRT) in the aeration tank resulting in decreased treatment efficiency.

The substrate feed point (primary clarifier effluent) to the aeration tank may be manipulated to partition the mixed liquor suspended solids (MLSS) into zones of high solids concentration and low solids concentration (Figure 15.3). With the zone of low solids concentration discharging to the secondary clarifier, the solids loading to the secondary clarifier is decreased and improved settleability occurs in the secondary clarifier. However, manipulating the substrate feed point results in decreased treatment efficiency.

A coagulant or metal salt such as lime ($Ca(OH)_2$) or a cationic polymer can be added to the secondary clarifier influent to improve settleability. Coagulants and polymers improve settleability by (1) increasing the floc particle density, (2) overcoming interfloc bridging and open floc formation, and (3) decreasing the surface area of the floc particle and filamentous bacteria. Coagulants add weight to the floc particles, while polymers remove large quantities of fine solids from the bulk solution.

The only rapid, nonspecific control measure that controls the growth of the filamentous bacteria is the addition of a toxicant. There are two toxicants that

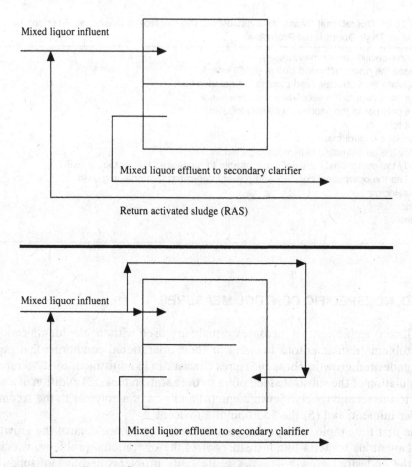

Mixed liquor influent

Mixed liquor effluent to secondary clarifier

Return activated sludge (RAS)

Mixed liquor influent

Mixed liquor effluent to secondary clarifier

Return activated sludge (RAS)

FIGURE 15.3 *Change in mixed liquor influent feed point to reduce solids loading to the secondary clarifier. By taking one-half of the influent flow (top) and discharging it to the rear portion of the aeration tank (bottom), a hydraulic gradient is established across the aeration tank. The hydraulic gradient results in the partitioning of the solids in the aeration tank into two zones. The first zone at the beginning of the tank contains a high concentration of MLVSS, while the second zone at the end of the tank contains a low concentration of MLVSS. Only the zone of low concentration of MLVSS is discharged to the secondary clarifier. This results in a decrease in solids loading upon the clarifier and reduced settleability problems.*

commonly are used for filamentous bacteria control. These toxicants are chlorine and hydrogen peroxide (H_2O_2). Chlorine is less expensive than hydrogen peroxide and usually is available on-site at wastewater treatment plants as a disinfectant for the final effluent.

Chlorine may be applied as gaseous chlorine (Cl_2) mixed with final effluent to produce free chlorine as hypochlorous acid (HOCl) and hypochlorous ion (OCl^-). Chlorine also may be applied as a solution of calcium hypochlorite ($Ca(OCl)_2$) or sodium hypochlorite (NaOCl). Chlorine may be introduced into the activated sludge process in (1) the RAS line, (2) a sidestream, (3) the aeration tank, and (4) mixed liquor effluent (Figure 15.4).

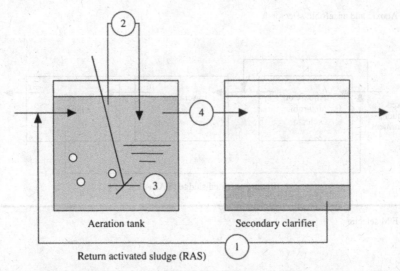

FIGURE 15.4 *Chlorine feed points for the control of undesired filamentous bacteria growth. There are four chlorine feed points. These points are (1) RAS, (2) side stream to and from the aeration tank, (3) directly to the aeration tank, and (4) mixed liquor effluent.*

The usual starting dose of chlorine for the control of undesired filamentous bacteria is 2–3 pounds of chlorine per 1000 pounds of mixed liquor volatile suspended solids (MLVSS) per day. To ensure the destruction of an adequate number of filamentous bacteria via chlorination and prevent the overchlorination of the activated sludge process, it is necessary to periodically monitor the impact of chlorine upon the biomass and the final effluent and adjust the dose of chlorine as needed.

SLOW, SPECIFIC CONTROL MEASURES

Slow, specific control measures for undesired filamentous growth consists of the following steps:

- Identify the undesired filamentous bacteria.
- Identify the operational conditions responsible for undesired filamentous growth.
- Adjust the operational condition to control the undesired filamentous growth.

SELECTORS

There are three selectors that commonly are used to control undesired filamentous growth. These selectors are anoxic, anaerobic, and F/M (Figure 15.5). The selectors have operational conditions that "select' or promote the growth of floc bacteria. Selectors are located upstream of the aeration tanks, and the RAS that contains the filamentous organisms is discharged to the selector.

Anoxic and anaerobic selectors

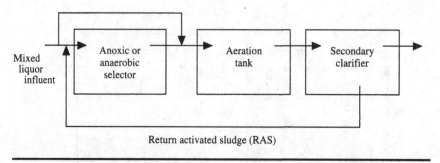

Return activated sludge (RAS)

F/M selector

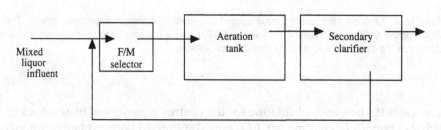

Return activated sludge (RAS)

FIGURE 15.5 *Selectors.*

An anoxic selector contains nitrate (NO_3^-) and/or nitrite (NO_2^-) and little or no free molecular oxygen. Some or all of the primary clarifier effluent that contains substrate or soluble cBOD is fed to the selector. In the presence of soluble cBOD, free molecular oxygen is quickly removed and an anoxic condition is established. The selector retention time is 1–2 hours; by recycling the RAS through the selector on a continuous basis, several filamentous bacteria that cannot use nitrate or nitrite can be controlled. These filamentous bacteria include *Haliscomenobacter hydrossis*, Nocardioforms, *Nosticoda limicola*, *Sphaerotilus natans*, type 1701, and type 021N.

An anaerobic selector is similar to an anoxic selector with the following exceptions: (1) the selector contains very little or no nitrate and/or nitrite and free molecular oxygen and (2) the selector has an anaerobic condition that permits the biological uptake of phosphorus. An anaerobic selector is capable of controlling aerobic filamentous bacteria and filamentous bacteria that are not capable of biological phosphorus uptake. Under anaerobic conditions, filamentous bacteria are not able to take up soluble substrate at a rate comparable to floc-forming, poly-P bacteria. The poly-P bacteria take up phosphorus under an aerobic condition (aeration tank immediately following the anaerobic selector) and use the stored phosphorus (energy) to outcompete filamentous bacteria for soluble substrate in the anaerobic selector.

An F/M selector has a retention time of only 5–15 minutes, and all of the primary effluent must be fed to the selector. Due to the short retention time in the selector, the selector may have an aerobic, anoxic, or anaerobic condition. Most F/M selectors have an aerobic condition.

An F/M selector establishes a substrate gradient. The gradient can be established by using an F/M gradient in the same tank or across several tanks. The F/M selector gives a substrate advantage to floc bacteria, because they absorb most of the soluble substrate at the influent end of the selector. An F/M selector can be used to control the undesired growth of low F/M filamentous bacteria.

Part V

Fermentation and Methane Production

16

Fermentative Bacteria

Organotrophic bacteria degrade organic compounds in order to obtain carbon and energy for cellular synthesis and activity. The degradation of organic compounds occurs intracellularly and results in the release of electrons from the hydrogen atoms in the organic compounds (Figure 16.1). The electrons provide the cell with energy and are removed from the cell by an electron transport molecule. The electron transport molecule may be free molecular oxygen (O_2), nitrate (NO_3^-), sulfate (SO_4^{2-}), carbon dioxide (CO_2), or an organic molecule (Table 16.1).

Bacterial cells can use only one electron transport molecule at any time, and the choice of the molecule always will be for the molecule that provides the cell with the most energy, provided that the following conditions are satisfied:

- The molecule is available for bacterial use.
- The bacterial cell has the enzymatic ability to use the molecule.

The degradation of organic molecules can occur with free molecular oxygen (aerobic degradation) or without free molecular oxygen (anaerobic degradation). If the degradation of an organic molecule—for example, glucose ($C_6H_{12}O_6$)—occurs with free molecular oxygen, the degradation is referred to as aerobic respiration (Equation 16.1). Aerobic respiration occurs in the aeration tank of an activated sludge process. Aerobic respiration results in the production of bacterial cells (sludge), carbon dioxide, and water.

$$C_6H_{12}O_6 + 6O_2 \rightarrow 6CO_2 + 6H_2O \tag{16.1}$$

There are four significant forms of anaerobic degradation of organic molecules that occur at wastewater treatment plants. These four are nitrate reduction

---

Wastewater Bacteria, by Michael H. Gerardi
Copyright © 2006 John Wiley & Sons, Inc.

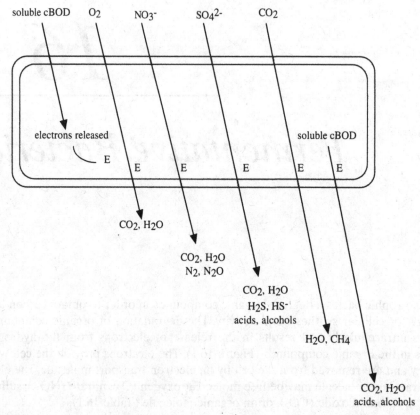

soluble cBOD O_2 NO_3^- SO_4^{2-} CO_2

electrons released   soluble cBOD

E   E   E   E   E   E

CO_2, H_2O

CO_2, H_2O
N_2, N_2O

CO_2, H_2O
H_2S, HS^-
acids, alcohols

H_2O, CH_4

CO_2, H_2O
acids, alcohols

FIGURE 16.1 *Molecules available for the removal of electrons released from the degradation of carbonaceous substrates. Depending upon the molecule used (O_2, NO_3^-, SO_4^{2-}, CO_2, or soluble cBOD) by bacteria to remove electrons (E) from the cell, a variety of substrates are produced.*

TABLE 16.1   Electron Transport Molecules Used in the Degradation of Soluble Substrate

Operational Condition	Transport Molecule	Biological Process
Aerobic	O_2	Aerobic degradation of substrate
Anoxic	NO_3^-, NO_2^-	Anoxic degradation of substrate
Anaerobic	SO_4^{2-}	Sulfate reduction and degradation of substrate
Anaerobic	CO_2	Methanogenesis
Anaerobic	Organic molecule	Fermentation and degradation of substrate

(denitrification), sulfate reduction, methanogenesis (methane production), and fermentation.

Facultative anaerobic bacteria perform nitrate reduction. Nitrate reduction or denitrification commonly occurs in a denitrification tank, an anoxic selector, and a secondary clarifier (Equation 16.2). The occurrence of denitrification as a secondary clarifier is commonly referred to as "clumping." Nitrate reduction results in the production of bacterial cells (sludge), carbon dioxide, water, and molecular nitrogen (N_2).

$$C_6H_{12}O_6 + 4NO_3^- \rightarrow 6CO_2 + 6H_2O + 2N_2 \tag{16.2}$$

Obligatory anaerobic bacteria perform sulfate reduction. Sulfate reduction typically occurs in an anaerobic digester. Sulfate reduction does occur in sewer systems. Sulfate reduction often occurs in a secondary clarifier and a thickener, if the settled solids remain too long in these treatment units and free molecular oxygen and nitrate are not present. Sulfate reduction results in the production of bacterial cells (sludge), carbon dioxide, water, sulfide (HS^-), and a variety of short chain organic compounds, mostly acids and alcohols. For example, from the anaerobic degradation of lactate ($CH_3CHOHCOOH$), acetate (CH_3COOH) is produced (Equation 16.3).

$$2CH_3CHOHCOOH + SO_4^{2-} + H^+ \rightarrow 2CH_3COO^- + 2CO_2 + 2H_2O + HS^- \tag{16.3}$$

Oxygen-intolerant, obligatory anaerobic bacteria perform methanogenesis. There are two major routes of methanogenesis that occur in an anaerobic digester. The major route of methane production is the "splitting" of acetate by aceticlastic, methane-forming bacteria (Equation 16.4). Splitting of acetate results in the production of bacterial cells (sludge), methane (CH_4), and water. The minor route of methane production is the reduction of carbon dioxide by hydrogen-oxidizing, methane-forming bacteria (Equation 16.5). The reduction of carbon dioxide results in the production of bacterial cells, methane, and water.

$$CH_3COO^- + H_2 \rightarrow CH_4 + H_2O + OH^- \tag{16.4}$$

$$4H_2 + CO_2 \rightarrow CH_4 + 2H_2O \tag{16.5}$$

Organic molecules undergo glycolysis as they are degraded. At the end of glycolysis two, three-carbon molecules of pyruvate ($CH_3COCOOH$) are produced. Pyruvate then is degraded further through aerobic respiration, nitrate reduction, sulfate reduction, methanogenesis, or fermentation.

Anaerobic degradation of organic compounds without free molecular oxygen, nitrate, sulfate, or carbon dioxide is fermentation. Fermentation results in the production of a variety of simplistic, soluble organic compounds, and the bacteria that perform fermentation are referred to as fermentative bacteria. Fermentation requires the use of an organic molecule to remove the electrons from the degrading compound. Fermentation is an inefficient process, and it releases little energy to the cell. Most of the energy released by the degraded compound remains in the fermented products. Fermentation typically occurs in an anaerobic digester, but it may also occur in sewer systems, a secondary clarifier, or a thickener.

Facultative anaerobic bacteria and strict anaerobic bacteria perform fermentation. Fermentation can occur by many different pathways (Figure 16.2). Fermentation results in the production of bacterial cells (sludge), water, and a large variety of organic compounds including lactic acid (Equation 16.6) and ethanol (Equation 16.7). Some fermentative pathways do produce carbon dioxide, while other pathways do not (Table 16.2). The types of organic compounds produced through fermentation are dependent upon the bacteria involved and the existing operational condition (e.g., pH and temperature). Fermentation results in the production of a

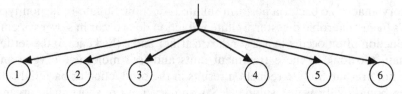

Substrate (glucose) degraded to pyruvate;
pyruvate fermented through several possible pathways,

Pathway #1, Acetone-butanol fermentation: sugar -- bacteria --- > products

    Bacteria: *Clostridium*

    Products: acetone, butanol, ethanol

Pathway #2, Butanediol fermentation: sugar -- bacteria --- > products

    Bacteria: *Aerobacter, Enterobacter, Erwinia, Klebsiella, Serratia*

    Products: acetate, 2,3-butanediol, butylene glycol, ethanol, carbon dioxide, hydrogen

Pathway #3, Butyric-butylic fermentation: sugar -- bacteria --- > products

    Bacteria: *Clostridium*

    Products: acetate, butyrate, ethanol, isopropanol, carbon dioxide, hydrogen

Pathway #4, Homolactic fermentation: sugar -- bacteria --- > products

    Bacteria: *Bacillus, Lactobacillus, Streptococcus*

    Products: lactate

Pathway #5, Mixed acid fermentation: sugar -- bacteria --- > products

    Bacteria: *Escherichia, Proteus, Providencia, Salmonella, Shigella, Yersinia*

    Products: acetate, ethanol, formate, lactate, succinate, carbon dioxide, hydrogen

Pathway #6, Propionic fermentation: sugar -- bacteria --- > products

    Bacteria: *Propionobacterium*

    Products: acetate, propionate, carbon dioxide

FIGURE 16.2 *Some major bacterial fermentative pathways. When soluble cBOD is degraded under anaerobic/fermentative conditions, a variety of products can be produced depending upon the bacteria involved, operational conditions present, and fermentative pathway used by the bacteria.*

variety of acids and alcohols, and it often is called "mixed acid and mixed alcohol" production. Due to the production of acids, fermentative bacteria also are known as acid-forming bacteria. Some acid-forming bacteria produce copious quantities of acetate and are known as acetogenic bacteria. Acetate is the major substrate for methane production in an anaerobic digester.

$$C_6H_{12}O_6 \rightarrow 2CH_3CHOCOOH \qquad (16.6)$$

TABLE 16.2   Significant Bacterial Fermentative Pathways

Fermentative Pathway	Products	Representative Bacterial Genus
Acetone–butanol	Acetone, butanol, ethanol,	*Clostridium*
Butanediol	Acetate,2,3-butanediol, butylene, ethanol, gylcol, lactate, CO_2, H_2	*Enterobacter*
Butyrate	Acetate, butyrate, CO_2, H_2	*Clostridium*
Lactate	Lactate	*Lactobacillus*
Mixed acid	Acetate, ethanol, lactate, CO_2, H_2	*Escherichia*
Propionate	Propionate	*Propionibacterium*

TABLE 16.3   Significant Genera of Fermentative Bacteria

Aeromonas	*Lactobacillus*
Bacteroides	*Pasteurella*
Bifidobacteria	*Propionobacterium*
Citrobacter	*Proteus*
Clostridium	*Providencia*
Enterobacter	*Salmonella*
Erwinia	*Serratia*
Escherichia	*Shigella*
Klebsiella	

$$C_6H_{12}O_6 \rightarrow 2CH_3CH_2OH + 2CO_2 \qquad (16.7)$$

In many of the fermentative pathways, hydrogen (H_2) is produced. The production of hydrogen gas is important in anaerobic digesters. First, hydrogen is a significant substrate for the production of methane (CH_4). Second, if hydrogen is not reduced to a low pressure in an anaerobic digester, the hydrogen pressure inhibits acetogenic bacteria.

Although facultative anaerobic bacteria and anaerobic bacteria (Table 16.3) are capable of fermentation, the more important bacteria are the strict anaerobic bacteria such as *Bacteroides*, *Bifidobacteria*, and *Clostridium*. These three genera of bacteria enter wastewater treatment plants through inflow and infiltration as soil and water organisms. Other fermentative bacteria enter wastewater treatment plants through inflow and infiltration and fecal waste.

There are two important groups of fermentative bacteria: the acidogenic bacteria and the acetogenic bacteria. The acidogenic bacteria or acid-formers such as *Clostridium* convert simple sugars, amino acids, and fatty acids to (1) organic acids such as acetate, butyrate, formate, lactate, and succinate, (2) alcohols such as ethanol and methanol, (3) acetone, and (4) carbon dioxide, hydrogen, and water. Several of these compounds are volatile and malodorous. Several of these compounds can be used directly by methane-forming bacteria, while other compounds can be converted to compounds that can be used by methane-forming bacteria.

Acetogenic bacteria produce acetate and hydrogen that can be used directly by methane-forming bacteria. Acetogenic bacteria convert several of the fatty acids [such as butyrate (Equation 16.8) and propionate (Equation 16.9)] and alcohols [such as ethanol (Equation 16.10)] that are produced by the acidogenic bacteria to acetate, hydrogen, and carbon dioxide.

$$CH_3CH_2CH_2COOH + 2H_2O \rightarrow 2CH_3COOH + 2H^+ \tag{16.8}$$

$$CH_3CH_2COOH + 2H_2O \rightarrow CH_3COOH + CO_2 + 3H_2 \tag{16.9}$$

$$CH_3CH_2OH + H_2O \rightarrow CH_3COOH + 2H_2 \tag{16.10}$$

Fermentation in wastewater treatment systems is involved in several significant operational conditions. These conditions are

- Production of malodorous organic compounds
- Production of PHB granules that are necessary for biological phosphorus removal
- Production of substrate for methane-forming bacteria in anaerobic digesters
- Rapid and undesired growth of septic-loving filamentous organisms including type 021N, type 0041, *Beggiatoa* spp., *Nosticoda limicola*, and *Thiothrix* spp.
- Rising sludge in secondary clarifiers and thickeners

17

Methane-Forming Bacteria

Methane-forming bacteria or methanogens are a specialized group of Archaea that utilize a limited number of substrates (Table 17.1), principally acetate (CH_3COOH), carbon dioxide, and hydrogen for methane production or methanogenesis. These substrates are the end products of more complex substrates that were degraded by fermentative bacteria.

Methane-forming bacteria are some of the oldest bacteria and are grouped in the domain Archaebacteria. The term "arachae" means ancient. Methane-forming bacteria have many shapes (bacillus, coccus, and spirillum), sizes (0.1 to 15 μm), and growth patterns (individual cells, filamentous chains, cubes, and sarcina). Methane-forming bacteria are oxygen-sensitive anaerobes and are found in habitats that are rich in degradable organic compounds. In these habitats oxygen is rapidly removed by bacterial degradation of the organic compounds. Some methane-forming bacteria live as symbionts in animal digestive tracts.

Most methane-forming bacteria are active in two temperature ranges, the mesophilic range from 30°C to 35°C and the thermophilic range from 50°C to 60°C. At temperatures between 40°C and 50°C, nearly all methane-forming bacteria are inhibited. Fluctuations in temperature within the mesophilic and thermophilic ranges are not directly harmful to methane-forming bacteria. However, fluctuations in temperature may change the dominant fermentative bacteria that produce the substrates that are used by methane-forming bacteria. If the dominant fermentative bacteria produce substrates that cannot be used by methane-forming bacteria, then methanogenesis is inhibited. Therefore, fluctuations in the operating temperature of anaerobic digesters should be minimal.

Methane-forming bacteria are active within the pH range of 6.8 to 7.2. Methane-forming bacteria are sensitive to pH values <6.8 and >7.2. With decreasing pH, methane-forming bacteria become less active, while fermentative bacteria remain

TABLE 17.1   Substrates Used by Methane-Forming Bacteria

Substrate	Chemical Formula
Acetate	CH_3COOH
Carbon dioxide	CO_2
Carbon monoxide	CO
Formate	$HCOOH$
Hydrogen	H_2
Methanol	CH_3OH
Methylamine	CH_3NH_2

active and continue to produce fatty acids. These acids destroy alkalinity and depress pH resulting in inhibition of methane-forming bacteria. Also, with decreasing pH, increases in the quantities of hydrogen sulfide (H_2S) and hydrogen cyanide (HCN) occur. These two inorganic compounds are highly toxic to methane-forming bacteria. With increasing pH, an increase in the quantity of ammonia (NH_3) occurs. Ammonia also is toxic to methane-forming bacteria. Therefore, anaerobic digesters should be operated at a near neutral pH value and should be monitored as needed to ensure an acceptable pH value and alkalinity residual.

Sufficient alkalinity is necessary for proper pH control. Alkalinity serves as a buffer that prevents rapid change in pH. Enzymatic activity of methane-forming bacteria is adversely affected by pH values <6.8 and >7.2. Adequate alkalinity in an anaerobic digester can be maintained by providing an acceptable volatile acid-to-alkalinity ratio. The range of acceptable volatile acid-to-alkalinity ratios is 0.1 to 0.2.

Because methane-forming bacteria reproduce very slowly (generation times of 3–30 days) and produce very few offspring (sludge) from the degradation of substrates (approximately 0.02 pounds of sludge per pound of substrate degraded), methane-forming bacteria require smaller quantities of most nutrients. However, there are a few nutrients that are required by methane-forming bacteria in quantities two to five times greater than most other bacteria. These nutrients are cobalt, iron, nickel, and sulfur.

Methanogenesis occurs through three basic biochemical reactions that are mediated by three different groups of methane-forming bacteria (acteoclastic methanogens, hydrotrophic methanogens, and methyltrophic methanogens). Acetoclastic methanogens produce methane by "splitting' acetate (Equation 17.1). Hydrogenotrophic methanogens produce methane by combining hydrogen and carbon dioxide (Equation 17.2), and methyltrophic methanogens produce methane by removing methyl ($-CH_3$) groups from simple substrates (Equation 17.3). In anaerobic digesters, acetoclastic methane-forming bacteria produce most of the methane, while hydrotrophic methane-forming bacteria produce approximately 30% of all methane. Methyltrophic methane-forming bacteria produce a relatively small quantity of methane in anaerobic digesters.

$$\text{Acetate} \xrightarrow{\text{acetoclastic methane-forming bacteria}} CH_4 + CO_2 \tag{17.1}$$

$$H_2 + CO_2 \xrightarrow{\text{hydrogenotrophic methane-forming bacteria}} CH_4 + 2H_2O \tag{17.2}$$

$$\text{Methanol} \xrightarrow{\text{methyltrophic methane-forming bacteria}} 2CH_4 + 2H_2O \tag{17.3}$$

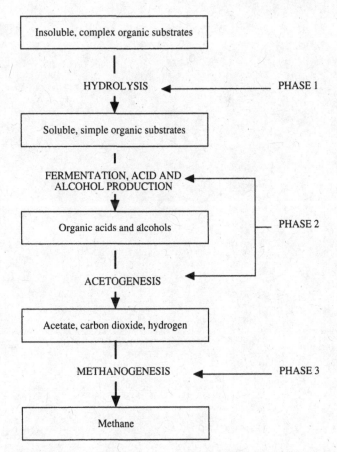

FIGURE 17.1 *Basic phases of anaerobic degradation of substrate.*

There are three basic biological events or phases that occur in municipal anaerobic digesters with respect to methane production. These phases are (1) hydrolysis, (2) fermentation and acetate production, and (3) methanogenesis (Figure 17.1). During hydrolysis, hydrolytic bacteria solubilize large and complex compounds to small and simple compounds. During fermentation, fermentative bacteria convert the newly formed soluble compounds to organic acids, alcohols, CO_2, and H_2. As part of fermentation, many of the acids and alcohols then are converted to acetate. During methanogenesis, methane-forming bacteria convert CO_2, H_2, acetate, and several other limited substrates to methane. A comprehensive review of methane-forming bacteria is provided in *The Microbiology of Anaerobic Digesters* in the Wastewater Microbiology Series.

Part VI

Toxicity

18

Septage

Septage is the liquid and solid (scum and sludge) waste produced in individual on-site wastewater disposal systems such as septic tanks and cesspools and is approximately 90–98% water. Septage periodically is removed from septic tanks and cesspools during cleaning operations and is often discharged to wastewater treatment plants for disposal and treatment.

Periodic removal or pumping of septage is essential to the long-term operation of septic tanks (Figure 18.1) and septic systems. Once removed, septage disposal occurs. Septage disposal practices vary across the nation and even within states. However, more uniform requirements for septage disposal practices are being adopted nationwide. Examples include the adoptions of new regulations (40 CFR503) in 1993 and federal guidelines for septic systems in 2002 by the U.S. Environmental Protection Agency.

There are several methods that are available for the disposal of septage. These methods incorporate the treatment of septage to control malodors, reduce solids volume, decrease the strength of pollutants, and destroy pathogens (disease-causing agents). The major methods available for the disposal of septage include (1) co-disposal with solid wastes, (2) co-treatment with wastewater, (3) land application, and (4) processing at separate facilities (Table 18.1).

For several reasons the co-treatment of septage with wastewater is a popular method for the disposal of septage (Table 18.2). This method is used at many wastewater treatment plants. However, small wastewater treatment plants are more vulnerable to "upset" conditions from the discharge of septage than large wastewater treatment plants, if septage is not properly discharged to the plants. Problems of organic overloading can be overcome by collecting and storing septage during the daytime hours and then discharging septage to the wastewater treatment plant when

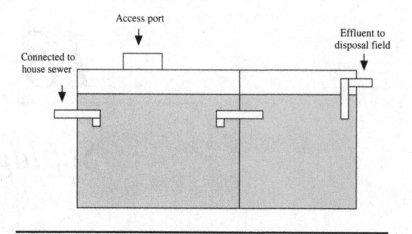

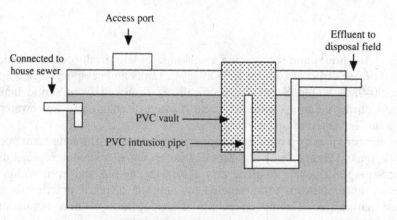

FIGURE 18.1 Septic tanks.

TABLE 18.1 Methods Available for the Disposal of Septage

Co-disposal with solid wastes
   Composting
   Landfilling
Co-treatment with wastewater
   Biological treatment
   Chemical treatment
Land application
   Subsurface application
   Surface application
Processing at separate facilities
   Biological treatment
   Chemical oxidation
   Composting
   Lime stabilization

TABLE 18.2   Reasons for the Co-treatment of Septage with Wastewater

Cooperation with regulatory agencies or political bodies
Only method available for disposal
Revenue for wastewater treatment plants
Use of excess treatment plant capacity
Year-round availability for septage haulers

TABLE 18.3   Significant Components in Domestic Wastewater and Septage

| Component | Range in Concentration (mg/liter) | | Typical Concentration (mg/liter) |
	Domestic Wastewater	Septage	Septage
BOD_5	110–400	2,000–6,000	3,000
COD	250–1,000	5,000–80,000	30,000
Grease	50–150	5,000–10,000	8,000
NH_4^+	12–50	100–800	400
Phosphorus	4–15	50–800	250
TKN	8–35	100–1,600	700
TSS	100–350	2,000–100,000	15,000
VSS	80–275	1,200–14,000	7,000

the influent organic loading is lowest. Biological treatment of septage at wastewater treatment plants usually is achieved in the activated sludge processes.

Septage is approximately 50 times as concentrated as domestic wastewater. Significant biological, chemical, and physical components of septage vary greatly (Table 18.3). Of the total nitrogen in septage, approximately 80% is organic nitrogen (TKN) and 20% is ionized ammonia (NH_4^+). Compared to those of wastewater, carbon (cBOD)-to-nutrient (nitrogen and phosphorus) ratios for septage are low, but not below levels that tend to inhibit acceptable degradation of cBOD. Concentrations of metals such as copper, lead, and zinc vary greatly in septage.

Variations in the concentrations of the components in septage occur for several reasons. The largest range in values of components often is found in communities that do not regulate the collection and disposal of septage. Variations in the concentration of metals and minerals in septage is due in large part to the potable water supply used and discharged to septic tanks.

Other components of septage that are of interest to wastewater treatment plant operators include grit, pathogens, foam, odor, and color. Septage contains relatively high levels of grit and pathogens, especially protozoan cysts and oocyts and helminth (worm) eggs. Septage is difficult to dewater and may foam easily upon agitation. Septage is malodorous, and the offensive odor associated with septage is produced through the anaerobic degradation of cBOD and the production and release of volatile fatty acids, nitrogen-containing compounds, and sulfur-containing compounds (Table 18.4). Color is a component of septage. Wastewater contaminated with septage becomes black.

The quantity of sludge and scum in septage also varies greatly and is affected by several factors. These factors include whether a kitchen food grinder (disposal unit) is used, the quantity of oil and grease discharged to the septic tank, and the

TABLE 18.4   Most Commonly Occurring Malodorous Compounds Produced through Anaerobic Degradation of BOD

Group	Compound	Chemical formula	Odor
Acid	Acetate	CH_3COOH	Vinegar
	Butyrate	$CH_2(CH_3)_2COOH$	Rancid
	Valeric acid	$CH_2(CH_3)_3COOH$	Sweat
Nitrogen-containing	Ammonia	NH_3	Irritating
	Dimethylamine	$(CH_3)_2NH$	Fish
	Indole	C_8H_6NH	Fecal
	Methylamine	CH_3NH_2	Rotten fish
	Skatole	C_9H_8NH	Fecal
Sulfur-containing	Ethylmercaptan	C_2H_5SH	Rotten cabbage
	Hydrogen sulfide	H_2S	Rotten egg
	Methylmercaptan	CH_3SH	Cabbage
	Methylsulfide	$(CH_3)_2S$	Rotten vegetables
	Methyldisulfide	$(CH_3)_2S_2$	Putrefaction

TABLE 18.5   Significant Components of a Properly Designed Septage Storage Tank

Adequate storage capacity to compensate for potential treatment plant problems
All-weather access for easy unloading of septage
Design considerations for minimization of malodor release
Fencing to prevent unauthorized entrance
Landscaping to reduce the visual impact of the storage tank
Safety signs to indicate potential danger of drowning and hazardous gases
Water-tight construction to prevent septage leaks

frequency of septage tank pumping. The frequency of pumping influences the degree of bacterial digestion of solids within the septic tank. The longer the digestion time that is provided, the smaller the quantity of sludge and scum that is produced.

Septage should be safely discharged to a wastewater treatment plant. Usually, septage is added to the influent wastewater or sludge. When added to the influent wastewater, septage should be slowly blended with the wastewater to minimize its impact upon the treatment process. When added to sludge, septage should be screened and slowly added to sludge for stabilization and dewatering. Whenever septage is added to wastewater influent or sludge, the feed rate of septage should be controlled.

Septage may be added to a wastewater treatment plant by discharging the septage through a bar screen to a receiving tank (Table 18.5). The septage is then slowly discharged to the treatment plant from the receiving tank at a constant rate. At some wastewater treatment plants, septage may be slowly discharged to a designated manhole. As the septage travels through the sewer system, it is mixed and is diluted with raw wastewater. The daily volume of septage received at a wastewater treatment plant should be closely regulated to prevent overloading of the treatment process. Limiting the hours for receiving septage can help to prevent overloading.

Problems associated with overloading from septage often occur at small wastewater treatment plants with limited capacity. These problems can be overcome by collecting and storing the septage during the daytime hours and then discharging the septage when loading conditions are favorable.

Receiving tanks often experience problems with spills, debris, and odors. To overcome these problems, a water supply with a hose should be provided at the receiving tank to clean the septage truck unloading area. A dumpster may be provided for the disposal of debris.

Malodor control can be achieved by increasing the pH of the septage, adding appropriate bioaugmentation products (bacterial cultures), or treating septage with a strong oxidant. The pH of septage may be increased with the use of hydrated lime (calcium hydroxide, $Ca(OH)_2$) or quicklime (calcium oxide, CaO). Cultures of *Pseudomonas* may be added to septage to control malodors. Species of *Pseudomonas* easily degrade sulfur-containing compounds that are associated with malodor production.

The addition of a strong oxidant to septage can not only decrease malodors but also reduce many of the undesired features of septage. Oxidants commonly added to septage include sodium hypochlorite ($NaOCl$), hydrogen peroxide (H_2O_2), ozone (O_3), and potassium permanganate ($KMnO_4$). In some cases the addition of a strong oxidant may enhance malodor production.

The addition of a strong oxidant may reduce the biochemical oxygen demand (BOD) and chemical oxygen demand (COD) associated with septage. The oxidant may promote biological degradation of some refractory organic compounds in septage, improve settleability and dewaterability of septage, and destroy pathogens. Also, nitrite may be oxidized to nitrate, and sulfides may be oxidized to sulfate or elemental sulfur.

Strong oxidants break some bonds in large complex molecules to form smaller, more easily degradable molecules and eliminate many large water-bound colloidal molecules. These actions improve biological degradation of organic compounds and improve solids settleability and dewaterability.

Unless hydrogen sulfide (H_2S) is present in relatively high quantities, it does not contribute to a malodorous condition. Sulfides (HS^-) combine with the metals in septage solids to form insoluble metallic sulfides.

The source of septage also is a concern to wastewater treatment plant operators. Septage should not, but may, contain industrial wastes, grease-trap wastes, or toxic wastes, and the wastewater treatment plant may not be prepared to handle these wastes. Therefore, the wastewater treatment plant should have a contract with all septage haulers who are approved to discharge septage to the wastewater treatment plant or sewer system (Table 18.6).

TABLE 18.6 Suggested Septage Hauler Requirements

Apply for and be issued a septage discharge permit prior to discharge and/or use of the wastewater treatment plant or sewer system.

Have proof of liability insurance with coverage limits as required.

Possess a valid septage hauling permit.

Provide indemnity bond, deposit, or other payment guarantee sufficient to guarantee payment fees.

Record and submit the source of septage discharged to the wastewater treatment plant or sewer system.

Wastes shall only be accepted for treatment, if the wastewater treatment plant processes and final effluent are not adversely affected.

The discharge and treatment of septage at wastewater treatment plants presents several operational concerns. These concerns include increased operational costs, release of malodors, growth of undesired filamentous organisms, and unacceptable wastes, especially toxic wastes.

Increased operational costs occur through increased sludge production and disposal and increased dissolved oxygen demand to satisfy the degradation of BOD. The degradation of BOD includes carbonaceous BOD and nitrogenous BOD. An appropriate discharge or treatment fee should be charged and collected from each septage hauler. This fee usually is based on gallons of septage discharged.

In order to protect the wastewater treatment plant against industrial wastes, grease-trap wastes, toxic wastes, or other unacceptable wastes, a small sample from each truckload of septage should be collected, labeled, and refrigerated for one to two weeks. If the wastewater treatment plant becomes upset after the discharge of septage to the treatment process, the collected samples can be analyzed to determine if the septage is responsible for the upset. A representative sample of septage may be obtained by collecting a small and flow-proportioned quantity of septage at the beginning, mid-point, and end of the septage flow stream to the receiving tank.

When discharged to an activated sludge process excess, quantities of sulfides and septic waste (short-chained acids and alcohols) promote the rapid and undesired growth of filamentous organisms such as *Beggiatoa* spp., *Thiothrix* spp., and type 021N. These organisms in undesired numbers contribute to settleability problems and loss of solids. When discharged to any biological wastewater treatment process, toxic wastes contribute to upset through inhibition of bacterial and protozoan activities.

19

Toxicity

Municipal wastewater treatment plants often treat a combination of industrial, commercial, and domestic wastewaters. Some municipal wastewater treatment plants also treat septage. These wastewaters and septage contain several significant components that are of concern to operators of municipal wastewater treatment plants (Table 19.1). Several of these components represent a risk of toxicity to aerobic, biological treatment units such as the activated sludge process and anaerobic, biological treatment units such as the anaerobic digester (Table 19.2). Although the discharge of toxic wastes in toxic amounts to municipal wastewater treatment systems is prohibited by Section 101(a) of the federal Clean Water Act, toxicity too often occurs in biological treatment units.

As National Pollution Discharge Elimination System (NPDES) permits become more restrictive for industrial wastewater discharges to municipal wastewater treatment plants, proposed chemical additions to the industrial wastewater may need to be monitored early to determine adverse impacts upon the biological processes of municipal wastewater treatment plants. Monitoring may need to include the toxic effects of an industrial wastewater.

Toxicity is the occurrence of an adverse impact upon the biomass in biological treatment units. By combining commercial, domestic, and industrial wastewaters in biological treatment units, the possibility occurs for the introduction of toxic wastes. Frequently, the presence of toxic wastes is sporadic, but the biomass of the treatment system can be seriously damage.

A toxic waste or toxicant is a compound or ion in the wastewater or sludge that has a deleterious effect on living organisms. Toxic wastes are known to cause toxicity or inhibition to cBOD removal and nBOD removal (nitrification) in the activated sludge process and cBOD removal and methane production in the

Wastewater Bacteria, by Michael H. Gerardi
Copyright © 2006 John Wiley & Sons, Inc.

TABLE 19.1   Significant Components of Municipal Wastewater

Component	Primary Source	Impact upon Biological Treatment Systems
Chelating agents	Industrial	Pass-through of heavy metals
Fats, oils, and grease	Commercial, domestic, and industrial	Foam production; Undesired growth of filaments; Toxicity in anaerobic systems
Heavy metals	Industrial	Toxicity
Oxygen demand, cBOD	Commercial, domestic, and industrial	Dissolved oxygen consumption; Sludge production
Oxygen demand, nBOD	Domestic, industrial	Dissolved oxygen consumption; Nitrification; Denitrification (clumping)
Pathogens	Domestic, industrial (slaughterhouses); I/I (cat, dog, and rodent wastes)	Disease transmission
Salts	Domestic (water softeners); Industrial	Increased salinity
Solvents	Industrial	Toxicity
Surfactants	Domestic, commercial, industrial	Foam production; Dispersion of biomass; Toxicity

anaerobic digester. There are two groups of toxic wastes: inorganic and organic (Table 19.3).

Inorganic compounds and ions **do not contain carbon (C) and hydrogen (H)** and may be placed into two broad categories. The first category contains the highly toxic compounds and ions such as arsenic (As), chromates, chromium (Cr), cyanide, and zinc (Zn). Many of these compounds and ions (arsenic, cadmium, chromium, copper, cyanide, mercury, and nickel) are classified as priority pollutants. Fluoride (F), another toxic anion, is commonly found in wastewater from electronics manufacturing facilities. The second category contains compounds and ions that are necessary for cellular growth but can be toxic in high concentrations. These compounds and ions, usually nontoxic, may induce changes in the metabolic processes that alter cellular growth patterns at high concentrations.

Organic compounds **do contain and hydrogen**. Among the toxic organic compounds are some that also are sources or carbon and energy (substrates) for some bacteria. Phenol and some phenolic compounds are examples of toxic organic compounds that also serve as substrates for some bacteria. Although phenol and phenolic compounds are toxic to many genera of bacteria, many species in the genus *Pseudomonas* can degrade these compounds and often proliferate in biological treatment units that receive phenol or phenolic compounds.

Significant toxic organic compounds include aromatic compounds, halogentated compounds, oils, lipophilic solvents, and anionic surfactants. Aromatic compounds of concern include benzene, toluene, and xylenes. Many of these compounds are highly toxic because they are non-ionic in charge or structure and easily dissolve in the cell wall of many bacteria.

Examples of toxic, halogenated aliphatic compounds include chloroform (CH_3Cl_3) or trichloromethane, carbon tetrachloride (CCl_4) or tetrachloromethane, 1,1,1-trichloroethane (CH_3CCl_3), and methylene chlorine (CH_2Cl_2) or dichloromethane. Chloroform is used as a solvent, particularly in lacquers, and in

TABLE 19.2 Threshold Concentrations (mg/liter) of Commonly Occurring Pollutants that Are Toxic to Biological Treatment Processes

Pollutant	Activated Sludge cBOD Removal	Activated Sludge nBOD Removal	Anaerobic Digester
Inorganic		Biological Treatment Process	
Ammonia		480	1500–3000
Arsenic	0.1		1.6
Cadmium	10–100		0.02
Chromium (Cr^{6+})	1–10	0.25	5–50
Chromium (Cr^{3+})	50		50–500
Copper	0.1–1	0.005–0.5	1–10
Cyanide	0.1–5	0.3	4
Iron	1000		5
Magnesium			1000
Lead	0.1	0.5	
Mercury	0.1–5		1300
Nickel	1–2.5	0.25	
Potassium			3500
Sodium			3500
Sulfate		500	
Sulfide	0.3	0.01	150
Zinc	0.3–1	0.01–0.5	5–20
Organic			
Acetone		840	
Benzene	100		
Carbon tetrachloride			10–20
Ceepryn (surfactant)	100		
Chloroform		10	10–16
Hydrazine		58	
Nacconol (surfactant)	200		
Phenol	50	4–10	
Skatole		16	
Toluene	200		

TABLE 19.3 Significant Toxic Inorganic Compounds and Ions and Toxic Organic Compounds

Group	Compound/Ion
Inorganic	Ammonia
	Chlorine
	Cyanide
	Heavy metals
	Sulfide
Organic	Halogenated compounds
	Oils
	Phenol and phenolic compounds
	Solvents
	Surfactants

the manufacturing of plastics. Carbon tetrachloride and 1,1,1-trichloroethane are used as degreasing agents. Methylene chlorine is used as a solvent for fats, oils, grease, and waxes.

Because neutrally charged (non-ionic) molecules move more easily and more quickly across cell membranes than charged (anionic and cationic) molecules,

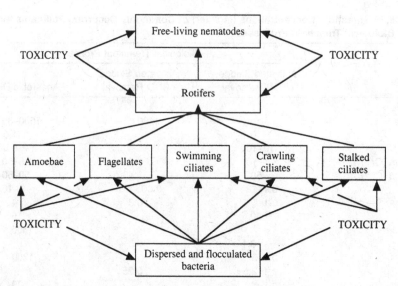

FIGURE 19.1   *Toxicity in the activated sludge process. With few exceptions, toxicity in biological treatment units such as the activated sludge process "attacks" all organisms.*

organic molecules produce a more rapid toxic impact in biological processes than inorganic molecules and ions. Because the cell wall of many bacteria (especially methane-forming bacteria) contains lipids, lipid-soluble or fat-soluble organic compounds easily dissolve in the cell wall and exert toxicity.

The undesired impacts of toxic wastes often are complex. The impacts include loss of treatment efficiency, permit violations, and increased operational costs. Usually, wastewater treatment plants are not designed for the undesired impacts of toxic wastes. Therefore, there is a need for operators to be (1) aware of potential toxic wastes, (2) able to monitor and detect the occurrence of toxicity and (3) able to make appropriate operational changes to prevent or reduce the impact of toxicity.

Often, the presence of toxic wastes in biological treatment units is sporadic. The discharge of toxic wastes in undesired quantities may be difficult to avoid, and the problem is complicated by the design of municipal plants to treat only nontoxic wastes. Although biological treatment units are susceptible to toxic wastes, the systems are resilient and have a large diversity of bacteria that may enable the system to tolerate or even degrade toxic wastes.

There are two terms that are used to express toxicity in biological treatment processes. These terms are "acute" toxicity and "chronic" toxicity. Acute toxicity is toxicity that is severe enough to damage a biomass within a relatively short period of time—that is, <48 hours. Chronic toxicity is toxicity that damages a biomass for a relatively long period of time.

Generally, toxic wastes are not specific; that is, the toxic wastes do not "attack" just one group of organisms. With exceptions, toxic wastes attack all organisms (Figures 19.1 to 19.3). However, some groups of organisms may be more susceptible to toxic wastes that other groups of organisms.

Toxic wastes attack bacteria, protozoa, and metazoa in activated sludge processes and bacteria in anaerobic digesters. The toxic attack upon bacterial cells results in

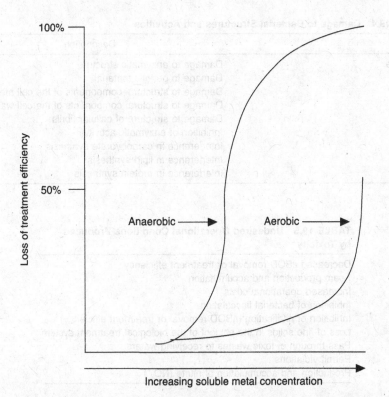

FIGURE 19.2 *Susceptibility of aerobic and anaerobic processes to the toxic effects of heavy metals. Because anaerobic bacteria in the digester obtain very little energy from the degradation of cBOD as compared to the aerobic and facultative anaerobic bacteria in the activated sludge process, anaerobic bacteria have very little energy available to repair damage caused by toxicity. Therefore, anaerobic digesters are much more "sensitive" to a toxic upset than the activated sludge process.*

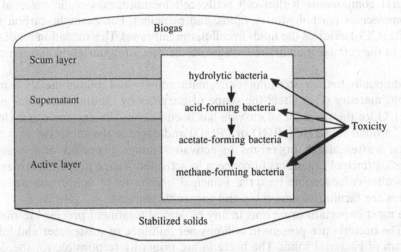

FIGURE 19.3 *Anaerobic digester toxicity and methane-forming bacteria. Methane-forming bacteria obtain very little energy from the production of methane. Compared to other bacteria in the anaerobic digester, methane-forming bacteria have the least amount of energy available to repair damage caused by toxicity and experience different forms of toxicity that other bacteria do not.*

TABLE 19.4   Damage to Bacterial Structures and Activities

Damage	Description
Structure	Damage to enzymatic structure
	Damage to genetic material
	Damage to structural components of the cell membrane
	Damage to structural components of the cell wall
	Damage to structure of cellular fibrils
Activity	Inhibition of enzymatic activity
	Interference in carbohydrate synthesis
	Interference in lipid synthesis
	Interference in protein synthesis

TABLE 19.5   Undesired Operational Conditions Produced by Toxicity

Decreased cBOD removal or treatment efficiency
Foam production and accumulation
Increased operational costs
Inhibition of bacterial flocculation
Inhibition of nitrification (nBOD removal or treatment efficiency)
Loss of fine solids in the effluent of the biological treatment system
Pass-through of toxic wastes to receiving waters
Permit violations
Production and accumulation of nitrite (NO_2^-)

damage to bacterial structures or essential bacterial activities (Table 19.4). The resulting damage produces undesired operational conditions (Table 19.5).

The earliest undesired impacts of toxicity upon biological treatment systems following exposure to toxicity occurs at the cellular level. Here, toxic wastes damage structural components within cell walls, cell membranes, genetic material, and macromolecules (carbohydrates, lipids, and proteins). For example, carbon tetrachloride (CCl_4) oxidizes the lipids in cellular membranes. This oxidation hinders the ability of the cellular membrane to regulate the flow of materials in and out of the cell.

Additionally, toxic wastes inhibit enzymatic activity and cellular metabolism. For example, mercury (Hg) alters the shape of enzymes by binding with thiol groups (—SH). Once the shape of an enzyme has been altered, the enzyme can no longer "wrap" around substrate (cBOD or nBOD) and degrade the substrate.

Toxic wastes attack organisms in activated sludge processes and anaerobic digesters. Principal organisms of concern in activated sludge processes are aerobic and facultative anaerobic bacteria. Principal organisms of concern in anaerobic digesters are facultative anaerobic and anaerobic bacteria.

The most important organisms in any biological treatment process are the bacteria. The bacteria are present in millions per milliliter of wastewater and billions per gram of bacterial solids. The bacteria are primarily responsible for the degradation of cBOD, degradation of nBOD, and removal of most fine solids—colloids, dispersed growth, and particulate material. The bacteria also are responsible for degrading or removing toxic wastes.

TABLE 19.6   Generation Times, Sludge Yields, and Energy Rank of Significant Bacterial Groups

Group	Generation Time (Approximate)	Sludge Yield (#VSS/#BOD Degraded)	Energy Available for Cellular Repair (Rank)
Organotrophs, activated sludge	30 minutes	0.6	1
Chemolithoautotrophs, activated sludge	2–3 days	0.06	2
Methane-forming bacteria	3–30 days	0.02	3

There are two significant groups of bacteria with respect to the degradation of cBOD and nBOD. These groups are the organotrophs and the chemolithoautotrophs. Organotrophs degrade organic compounds or cBOD in activated sludge processes and anaerobic digesters. Bacteria that degrade the cBOD in activated sludge processes include aerobic and facultative anaerobic bacteria. Bacteria that degrade the cBOD in anaerobic digesters include facultative anaerobic and anaerobic bacteria. A major group of anaerobic bacteria found in anaerobic digesters is the methane-forming bacteria. Chemolithoautotrophs or nitrifying bacteria degrade the ionized ammonia (NH_4^+) and nitrite (NO_2^-) in activated sludge processes.

Organotrophs and chemolithoautotrophs degrade cBOD and nBOD, respectively, to obtain energy for cellular activity including growth and cellular repair. However, there is a significant difference in the quantity of energy obtained by these bacteria from the degradation of their respective BOD. Organotrophs obtain far more energy than chemolithoautotrophs when they degrade equivalent quantities of substrate.

Organotrophs and chemolithoautotrophs obtain more energy from the degradation of cBOD and nBOD, respectively, in the activated sludge process than methane-producing bacteria obtain from the degradation of cBOD in the anaerobic digester when these bacteria degrade equivalent quantities of substrate. With more energy available for growth and cellular repair, the activated sludge bacteria reproduce more frequently and in larger numbers than methane-forming bacteria (Table 19.6).

Because bacteria in the activated sludge process obtain more energy than methane-forming bacteria from the degradation of their respective substrates, the activated sludge bacteria can repair cellular damage caused by toxic wastes more often and more quickly. Therefore, activated sludge processes are more tolerant (less susceptible) to toxic wastes (e.g., heavy metals) than anaerobic digesters.

In the activated sludge process, organotrophs are more tolerant of toxic wastes than chemolithoautotrophs (nitrifying bacteria). For example, the minimum concentration of some heavy metals tolerated by organotrophs often are 10–50 times greater than those tolerated by nitrifying bacteria (Table 19.7).

FORMS OF TOXICITY

There are some toxic wastes (heavy metals) that harm most or all organisms in biological treatment processes. Also, there are some toxic wastes that are unique and

TABLE 19.7 Examples of Minimum Concentrations of Heavy Metals that Inhibit cBOD Removal and nBOD Removal

Metal	Inhibitory Concentration (mg/l)	
	cBOD Removal	nBOD Removal
Copper (Cu)	0.1	0.005
Nickel (Ni)	1	0.25
Zinc (Zn)	0.3	0.01

TABLE 19.8 Toxic Wastes Unique to Nitrifying Bacteria

Recognizable, soluble cBOD
Substrate toxicity

TABLE 19.9 Toxic Wastes Unique to Methane-Forming Bacteria

Alternate electron acceptors
Long-chain fatty acids

harm only a group of organisms. These unique toxic wastes affect nitrifying bacteria (Table 19.8) and methane-forming bacteria (Table 19.9).

TOXICITY AND NITRIFYING BACTERIA

Because the nitrifying bacteria, *Nitrosomonas*, *Nitrosospira*, *Nitrobacter*, and *Nitrospira*, obtain very little energy from the oxidation of ionized ammonia (NH_4^+) and nitrite (NO_2^-), these organisms have little energy available to reproduce or repair cellular damage as compared to organotrophic bacteria. Nitrifying bacteria reproduce every 2–3 days and make up only 3–10% of the bacterial population as compared to organotrophic bacteria that reproduce every 30 minutes and make up over 90% of the bacterial population.

Due to their small numbers and limited quantity of energy available to repair cellular damage caused by toxicity, nitrifying bacteria are more susceptible to toxicity than the organotrophic bacteria in the activated sludge process. This susceptibility means that nitrifying bacteria experience toxicity at lower concentrations of toxic wastes than do bacteria that degrade cBOD. This susceptibility also means that nitrifying bacteria cannot repair as quickly and as often the cellular damage caused by toxicity.

In addition to increased susceptibility to toxicity, nitrifying bacteria also experience two unique forms of toxicity. These forms of toxicity are recognizable, soluble cBOD and substrate.

Recognizable, Soluble cBOD

It is generally accepted that nitrification can occur when the cBOD in the biological process is <40mg/liter. When the cBOD has been reduced to this relatively low

TABLE 19.10   Examples of Recognizable, Soluble cBOD

Compound	Formula	Carbon Units
Aminoethanol	$CH_3NH_2CH_2OH$	2
n-Butanol	$CH_3CH_2CH_2CHOH$	4
t-Butanol	$(CH_3)_3COH$	4
Ethanol	CH_3CH_2OH	2
Ethyl acetate	$CH_3CO_2C_2H_5$	4
Methanol	CH_3OH	1
Methylamine	CH_3NH_2	1
n-Propanol	$CH_3CH_2CH_2OH$	3
t-Propanol	$(CH_3)_2CHOH$	3

concentration, nitrification occurs for two reasons. First, nitrifying bacteria are able to compete successfully for dissolved oxygen as the cBOD concentration decreases and organotrophic bacteria require less dissolved oxygen. Second, those forms of cBOD that inhibit enzymatic activity in nitrifying bacteria are either degraded completely or degraded to a concentration that does not inhibit nitrification. The forms of cBOD that inhibit nitrifying bacteria are highly soluble and simplistic in structure (Table 19.10). These forms of cBOD easily enter the cells of nitrifying bacteria and are "recognized" by their enzyme systems—that is, bind to and inhibit enzymes.

Substrate Toxicity

Substrate toxicity occurs when the concentration of ionized ammonia or nitrite becomes excessive in the biological treatment system. With increasing pH, ionized ammonia is converted to free ammonia (NH_3) (Equation 19.1). Free ammonia is toxic to nitrifying bacteria. With decreasing pH, nitrite is converted to free nitrous acid (HNO_2) (Equation 19.2). Free nitrous acid is highly toxic to nitrifying bacteria. Nitrite is produced during nitrification when ionized ammonia is oxidized. If nitrification proceeds properly, the nitrite is oxidized quickly to nitrate (NO_3^-).

$$NH_4^+ + OH^- \rightarrow NH_3 + H_2O \qquad (19.1)$$

$$NO_2^- + H^+ \rightarrow HNO_2 \qquad (19.2)$$

Most activated sludge processes that nitrify do so at near-neutral pH values (6.8 to 7.2). Although nitrifying bacteria can oxidize nBOD at pH values >7.2, significant changes in pH (i.e., >0.3 standard units per day) adversely affect nitrifying bacteria. The adverse effect results in incomplete nitrification and often the production and accumulation of nitrite.

TOXICITY AND METHANE-FORMING BACTERIA

Methane-forming bacteria are the most susceptible bacteria to toxicity in the anaerobic digester. When strict anaerobic, methane-forming bacteria are compared to

TABLE 19.11 Forms of Toxicity to Methanogens

Form	Description or Example
Alkali/alkaline metals	Ca^{2+}, K^+
Ammonia	NH_3
Cyanide	HCN or cyano-compounds
Feedback inhibition	H_2, volatile fatty acids
Heavy metals	Cu^{2+}, Zn^{2+}
Long-chain fatty acids	Carprylic acid, lauric acid
Sulfate/nitrate (alternate electron acceptors)	SO_4^{2-}, NO_3^-
Tannins	Phenolic compounds

TABLE 19.12 Long-Chain Fatty Acids that Are Toxic to Methanogens

Fatty Acid	Number of Carbon Units
Caprylic	8
Capric	10
Lauric	12
Myristic	14
Oleic	16

facultative anaerobic bacteria and other anaerobic bacteria that perform hydrolysis, fermentation, and sulfate reduction in the anaerobic digester, the susceptibility of methane-forming bacteria to toxicity can be illustrated.

First, methane-forming bacteria are found in smaller numbers as compared to facultative anaerobic bacteria and other anaerobic bacteria in the anaerobic digester. Second, methane-forming bacteria obtain very little energy from the degradation of their substrates as compared to the energy obtained by facultative anaerobic bacteria and other anaerobic bacteria from the degradation of their respective substrates.

The small population size of the methane-forming bacteria and the small amount of energy available to repair damage caused by toxicity make the methane-producing bacteria and the anaerobic digester very susceptible to toxicity at relatively low concentrations of many toxic wastes (Table 19.11). In addition, their unique lipid structure of the cell wall makes the methane-forming bacteria susceptible to toxicity from long-chain fatty acids (Table 19.12).

Akali/Alkaline (Earth) Metals

There are four alkali metals of concern with respect to anaerobic digester toxicity. These metals are calcium (Ca^{2+}), magnesium (Mg^{2+}), potassium (K^+), and sodium (Na^+). Of these metals, three influence salinity and affect the structure of the cell membrane and its ability to regulate the flow of substrate and nutrients into the cell and wastes from the cell. The three metals that influence salinity are magnesium, potassium, and sodium.

Small concentrations of these four alkali metals are stimulatory to bacterial activity, while high concentrations are inhibitory to bacterial activity. The inhibitory effects of the alkali metals can be controlled by diluting the metals or adding antagonistic elements (e.g., potassium for sodium).

Ammonia

Ionized ammonia (NH_4^+) is not toxic and is produced in the anaerobic digester through the degradation of amino acids, proteins, polymers, and surfactants. Ammonia (NH_3) is toxic and is produced in the anaerobic digester with increasing pH (Equation 19.3).

$$NH_4^+ + OH^- \rightarrow NH_3 + H_2O \qquad (19.3)$$

Ammonia is toxic to methane-forming bacteria at concentrations of 1500–3000 mg/liter. Ammonia toxicity in an anaerobic digester can be controlled by decreasing pH and diluting the digester feed sludge. Ammonia toxicity also can be controlled by slowly acclimating the biomass.

Cyanide

Although hydrogen cyanide (HCN) could be considered an organic compound, it is an example of a highly toxic inorganic compound. Hydrogen cyanide and cyanide-containing (cyano) compounds are found in wastewaters from chemical manufacturers, coking operations, electroplating operations, metallurgical operations, and petrochemical operations. Although cyanide toxicity is concentration- and time-dependent, many bacteria including methane-forming bacteria are capable of recovery from this toxicant. Many bacteria can acclimate to cyanide, and cyanide is not always toxic. Cyanide hinders methane production from acetate but doesn't hinder methane production form carbon dioxide and methanol.

Cyanide also is produced biologically by cyanogenic plants including alfalfa, almonds, peaches, and sorghum, as well as by cyanogenic bacteria and fungi, under specific growth conditions. Several amino acids, particularly glycine, serve as precursors of cyanide. Cyanide is not only produced by cyanogenic bacteria but also detoxified, assimilated, and removed from wastewater, sludge, and soil by some organotrophic bacteria.

Feedback Inhibition

During the anaerobic degradation of organic substrates, several compounds are produced that can cause inhibition of methane-forming bacteria if these compounds accumulate to relatively high concentrations. These compounds include hydrogen (H_2) and volatile fatty acids. The accumulation of hydrogen results in an increase in hydrogen pressure. This pressure inhibits acetate production. Acetate is the primary substrate for methane-producing bacteria in an anaerobic digester. The accumulation of volatile fatty acids results in destruction of alkalinity and drop in pH.

Feedback inhibition occurs because the accumulation of hydrogen and volatile acids inhibits the natural step-by-step bacterial activity in the degradation of organic substrates. Control of feedback inhibition can be corrected by adding alkalinity and maintaining an acceptable volatile acid-to-alkalinity ratio.

Heavy Metals

Heavy metals can be toxic to methane-forming bacteria at relatively low concentrations. However, the metals must be in solution in order to exert toxicity. Fortunately, most metals transferred from primary sludge and secondary sludge to an anaerobic digester are bonded to solids and cannot enter bacterial cells and cause toxicity. Metals in solution that are transferred to an anaerobic digester usually are chelated by organic acids in the digester. Metals in solution that are not chelated can be precipitated as forms of sulfides and carbonates. When chelated or precipitated, the metals cannot enter bacterial cells and cannot cause toxicity.

Long-Chain Fatty Acids

Long-chain fatty acids can exert toxicity individually or collectively. Significant sources of long-chain fatty acids include wastewaters from homes, edible oil refinery processes, palm oil processing, slaughterhouses, and wool scouring processes. Methane-forming bacteria have unique cell wall structure. A significant component of the cell wall, especially in acetoclastic (acetate-degrading) methane-forming bacteria, is the lipid portion. Lipids in the cell wall are structurally similar to many long-chain fatty acids including caprylic, capric, lauric, myristic, and oleic.

Due to the similarities in structure between the lipids in the cell wall of the bacteria and the long-chain fatty acids, the fatty acids dissolved in the cell wall and cause toxicity. Toxicity from long-chain fatty acids can be controlled by monitoring and regulating significant discharges of these acids or by ensuring adequate degradation of the fatty acids upstream of the anaerobic digester.

Sulfide

Hydrogen sulfide (H_2S) is one of the most toxic compounds to methane-forming bacteria. Sulfide is produced in an anaerobic digester through the reduction of sulfate (SO_4^{2-}) and by the degradation of sulfur-containing compounds such as proteins. Only soluble sulfides are toxic. Toxicity occurs when the sulfide concentration in the anaerobic digester exceeds 150 mg/liter.

Because diffusion of compounds through the cell membrane of organisms is more rapid when compounds are non-ionized, sulfide toxicity is very pH-dependent. Hydrogen sulfide moves through the cell membrane more quickly than sulfide (HS^-). As pH drops (<7) in the anaerobic digester, the concentration of hydrogen sulfide increases while the concentration of sulfide decreases.

Within anaerobic digesters the acid-producing bacteria are more tolerant of hydrogen sulfide than methane-forming bacteria. Therefore, in the presence of toxic levels of hydrogen sulfide, methane production by methane-forming bacteria is inhibited while acid production by acid-forming bacteria continues. Without the conversion of the acids to methane, the acids accumulate, and the anaerobic digester may become "sour."

Sulfide toxicity can be minimized by (1) controlling the pH of the digester, (2) precipitating sulfide as a metal salt, (3) recycling the sludge (biomass) to select for sulfide-tolerant bacteria, and (4) diluting the digester feed sludge. Hydrogen sulfide also can be removed by stripping it from the digester through biogas scrubbing.

TABLE 19.13 Oxidation–Reduction Potential (ORP) and Bacterial Activity in an Anaerobic Digester

ORP (mV)	Bacteria Activity
+100 to −100	NO_3^- available and used to degrade cBOD; denitrification occurring
<−100	SO_4^{2-} available and used to degrade cBOD; sulfate reduction and acid production occurring
<−200	Fermentation and acid production occurring
<−300	Methane production occurring

Sulfate and Nitrate

Sulfate (SO_4^{2-}) is used by sulfate-reducing bacteria to degrade organic substrate. When organic substrate is degraded, the electrons released by the substrate are removed from the bacterial cell by sulfate. Because sulfate-reducing bacteria and methane-forming bacteria utilize many identical substrates, competition for substrates between these two bacterial groups occurs. Because sulfate-reducing bacteria are more active than methane-forming bacteria, they can better compete for the available substrates, especially when more electron acceptors are available (i.e., increase in sulfate concentration). Nitrate (NO_3^-) also acts as an alternate electron acceptor for many bacteria in the anaerobic digester.

The presence of sulfate as well as nitrate in digester feed sludge increases the oxidation–reduction potential (ORP) of the digester sludge (Table 19.13). An increase in ORP above −300 mV is inhibitory to methane-forming bacteria. An ORP value between −100 and −300 mV inhibits methane-forming bacteria and the production of methane but does not inhibit acid-forming bacteria.

Tannins

Tannins are phenolic compounds that are toxic to methane-forming bacteria. Tannins are produced and found in apples, bananas, beans, cereal, and coffee. It is suspected that tannins inhibit specific enzyme sites in methane-forming bacteria.

Additional Toxic Compounds

Additional compounds that are toxic to anaerobic digester bacteria, especially methane-forming bacteria, include chlorinated hydrocarbons such as chloroform, benzene ring compounds, and formaldehyde. Benzene ring compounds include benzene, pentachlorophenol, phenol, and toluene. Pentachlorophenol is the most toxic of the benzene ring compounds.

Indicators of Toxicity

The presence of toxic wastes in biological treatment units affects most or all organisms. Because toxicity affects cellular structure and cellular activity, there are several biological, chemical, and physical indicators of toxicity in the activated sludge process (Table 19.14) and the anaerobic digester (Table 19.15).

TABLE 19.14   Indicators of an Unstable Biomass in the Activated Sludge Process

Indicator	Decrease in Value	Increase in Value	Add Check Mark (√) if Appropriate
COD-to-BOD ratio		X	
Effluent ionized ammonia concentration		X	
Effluent BOD concentration		X	
Effluent conductivity		X	
Effluent nitrate concentration	X		
Effluent nitrite concentration		X	
Effluent orthophosphate concentration		X	
Effluent TSS concentration		X	
Favorable, dominant protozoan groups	X		
Mixed liquor dissolved oxygen concentration		X	
Protozoan/metazoan activity	X		
Protozoan/metazoan numbers	X		
Specific oxygen uptake rate (SOUR)	X		

TABLE 19.15   Indicators of an Unstable Biomass in the Anaerobic Digester

Indicator	Decrease in Value	Increase in Value	Add Check Mark (√) if Appropriate
Alkalinity concentration	X		
Biogas production	X		
Methane production	X		
Percent carbon dioxide in biogas		X	
pH	X		
Volatile acid concentration		X	
Volatile solids destruction	X		

Increase in Nutrient Concentrations in Reactor Effluent

Toxicity in the activated sludge process is one of several reasons for an unstable biomass that is responsible for increases in the quantities of ionized ammonia (NH_4^+) and orthophosphate ($H_2PO_4^-/HPO_4^{2-}$) in the filtrate of an aeration tank effluent (Table 19.16). Bacteria degrade BOD in order to obtain energy and carbon for cellular activity and growth (MLVSS). In order to maintain adequate cellular activity and growth, nitrogen and phosphorus in the form of ionized ammonia and orthophosphate, respectively, are removed from the bulk solution of the aeration tank during the degradation of BOD. However, during a toxic event in the aeration tank, decreased cellular activity occurs as a result of damage to bacterial enzyme systems.

Decreased enzymatic activity results in decreased degradation of BOD and decreased MLVSS production. With less MLVSS production occurring, decreased quantities of nutrients are removed from the bulk solution. Therefore, increased quantities of nutrients leave the aeration tank. The ionized ammonia and orthophosphate that are not used in the aeration tank can be found in the filtrate of the aeration tank effluent.

TABLE 19.16 Operational Conditions Responsible for Increased Concentrations of Ionized Ammonia and Orthophosphate in the Filtrate of an Aeration Tank

Operational Condition	Nutrient	
	Ionized Ammonia	Orthophosphate Ions
Decrease organic loading	X	X
Die-off of large numbers of bacteria (lack of substrate or endogenous respiration)	X	X
Excess ionized ammonia in influent	X	
Excess orthophosphorus in influent		X
Increased HRT (solubilization of nutrients)	X	X
Loss of nitrification	X	
Toxicity	X	X

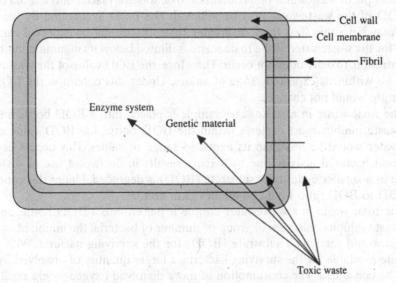

FIGURE 19.4 Toxic "attack" upon a bacterial cell. Depending upon the form of the toxic waste, toxicity "attacks" critical cellular structures and important cellular activities. Critical cellular structures attacked include the cell wall, cell membrane, fibrils, and genetic material. Important cellular activities attacked include enzymatic activity and the regulation of genetic materials.

Increase in Total Suspended Solids in Aeration Tank Effluent

Because toxicity weakens or destroys bacterial structures (e.g., cellular fibrils; Figure 19.4) and adversely affects ciliated protozoan and metazoan activity and numbers, increased total suspended solids (TSS) are found in the aeration tank effluent. The TSS consists of particulate materials and dispersed growth. A large quantity of colloids also is associated with the TSS.

Bacterial fibrils contribute to floc formation, by holding bacteria together. When bacteria are injured or die during a toxic occurrence, the fibrils are damaged and the floc particles become weak. Weakened floc particles are easily sheared under the existing turbulence levels of the activated sludge process, and these sheared floc particles release fine solids (colloids, dispersed cells, and particulate materials).

Bacterial fibrils not only hold bacteria together but also remove fine solids from the bulk solution through their adsorption to the ionized or active sites on the fibrils. Toxic wastes that damage the fibrils or neutralize the active sites result in the inability of the floc particles to remove large quantities of fine solids. Toxic wastes that neutralize the active sites of fibrils are heavy metals. The degree of ionization of the active sites is pH-dependent and is adversely affected at pH values greater than 8 and less than 6.5.

Chemical Oxygen Demand-to-Biochemical Oxygen Demand Ratio

Chemical oxygen demand (COD) is not affected by bacterial activity. Therefore, COD is not affected by toxicity. Biochemical oxygen demand (BOD) is affected by bacterial activity. However, BOD may or may not be affected by toxicity.

If a sample of wastewater that contains a toxic waste is placed into a BOD bottle, the BOD of the wastewater may be affected. The BOD as affected by the toxic waste may be within, less than, or greater than the normal or expected range of values for the wastewater. If the toxic waste is diluted below its minimum inhibitory concentration, toxicity does not occur. Therefore, the BOD value of the wastewater would be within its expected range of values. Under this condition the COD-to-BOD ratio would not change.

If the toxic waste in a wastewater sample is placed into a BOD bottle and the toxic waste inhibits most bacteria within the BOD bottle, the BOD value of the wastewater would be less than its expected range of values. This occurs because depressed bacterial activity due to toxicity results in decreased use of dissolved oxygen as a smaller quantity of substrate (BOD) is degraded. Under this condition the COD-to-BOD ratio would be greater than normal.

If the toxic waste in a wastewater sample is placed into a BOD bottle, and the toxic waste inhibits only a small group or number of bacteria, the inhibited or dead bacteria would serve as a substrate (BOD) for the surviving bacteria. With more substrate available for the surviving bacteria, a larger quantity of dissolved oxygen would be consumed. The consumption of more dissolved oxygen would result in a BOD value for the wastewater that would be higher than its expected range of values. Under this condition the COD-to-BOD ratio would be less than normal.

Conductivity

Conductivity is a measure of the ability of an aqueous solution to carry an electric current [ohm (Ω)]. Because the current is relatively small, the current is measured as a micro-ohm ($\mu\Omega$), and the current is measured over a distance of one centimeter, μohm/cm.

The ability of an aqueous solution to carry an electric current depends on the presence of ions in solution. Conductivity is influenced by the ions present (Table 19.17), their concentrations, their mobility, their charge or valence, and the temperature of the solution.

Aqueous solutions of most inorganic compounds are relatively good conductors of electric current, because these compounds dissociate easily in water. Aqueous solutions of most organic compounds are relatively poor conductors of electric current, because these compounds do not dissociate in water. Therefore, the pres-

TABLE 19.17 Relative Strength of Some Common Ions

Cations:
$H^+ > Fe^{3+} > Ca^{2+} > Fe^{2+} > Mg^{2+} > K^+ > NH_4^+ > Na^+$
Anions:
$OH^- > SO_4^{2-} > CO_3^{2-} > HPO_4^{2-} > Cl^- > NO_3^- > HCO_3^-$

TABLE 19.18 Conductivity Values for Common Water and Wastewater Samples

Sample	Value ($\mu\Omega$/cm)
Pure water	0
Distilled water	0.5–3
Potable water	50–1500
Domestic wastewater	600–800
Final effluent	600–900
Industrial wastewater (inorganic discharge)	>10,000

ence of increasing concentrations of dissociating inorganic compounds may result in a significant change from the expected range of conductivity values for water or wastewater (Table 19.18).

When toxicity occurs in a biological process, the conductivity value of its process may be significantly different from its typical range of values. This difference can occur from two factors. First, regardless of the type of toxic waste present in the biological process, bacteria that die from toxicity undergo lysis; that is, they break apart and release cellular wastes. Many of the wastes that are released during lysis contribute to conductivity. Second, if the toxic waste is ionic in structure or dissociates, the toxic waste also contributes to conductivity.

Microscopic Indicators

Protozoa and metazoa are two significant groups of "higher" life forms in the activated sludge process. They enter the process through inflow and infiltration (I/I) as soil and water organisms and make up approximately 5% of the weight of the MLVSS. Ciliated protozoa may be present as high as 50,000 per milliliter. Metazoa usually are present in highly variable numbers. Unless the MCRT of the activated sludge process is >28 days, most metazoa are not provided with sufficient time to reproduce and usually are present in the activated sludge process in relatively small numbers (<200 per milliliter).

Protozoa and metazoa often are used as bioindicators of the "health" of the activated sludge process or mixed liquor. Protozoa are microscopic, single-celled organisms that are animal-like, fungus-like, and plant-like. Protozoa in the mixed liquor process commonly are classified into five groups. These groups are amoebae, flagellates, free-swimming ciliates, crawling ciliates, and stalked ciliates.

Generally, amoebae and flagellates dominate the mixed liquor under adverse or harsh operational conditions including low dissolved oxygen concentration, toxicity, and low MCRT, while the ciliated protozoa, especially crawling and stalked

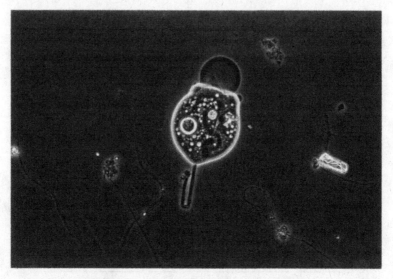

FIGURE 19.5   *Gassing of stalked ciliated protozoa.*

ciliates, dominate under favorable conditions, including high dissolved oxygen concentration, low pollution, and absence of toxicity. With improving quality of the operational conditions within the activated sludge process, progression in dominant protozoan groups occurs from amoebae to flagellates to free-swimming ciliates to crawling ciliates to stalked ciliates. With decreasing quality of the operational conditions within the activated sludge process, regression in dominant protozoan groups occurs from stalked ciliates to crawling ciliates to free-swimming ciliates to flagellates to amoebae.

Metazoa are multicellular animals that may be microscopic or macroscopic in size. Commonly observed metazoa in the activated sludge process include bristleworms, flatworms, free-living nematodes, rotifers, and waterbears. The most commonly observed metazoa are free-living nematodes and rotifers.

Because protozoa and metazoa are observed and counted easily during microscopic examinations of the mixed liquor, their numbers, activity, structure, and changes in dominant and recessive groups are often used as bioindicators of the presence of adverse or harsh operational conditions, including the presence of toxic wastes.

The presence of toxicity within the activated sludge process may be associated with the following microscopic bioindicators:

- Decrease in activity or loss of activity
- Decrease in numbers
- Regression in dominant protozoan groups
- Gassing or bubble production in stalked ciliated protozoa (Figure 19.5)

If the toxic wastes are surfactants, dispersion of rotifers (Figures 19.6 and 19.7) and free-living nematodes (Figures 19.8 and 19.9) and rotifers may occur.

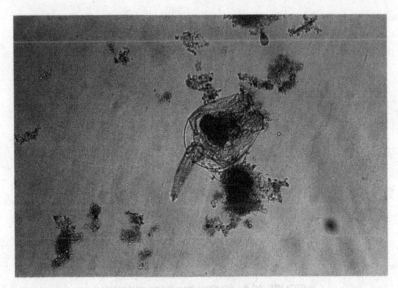

FIGURE 19.6   Healthy rotifer.

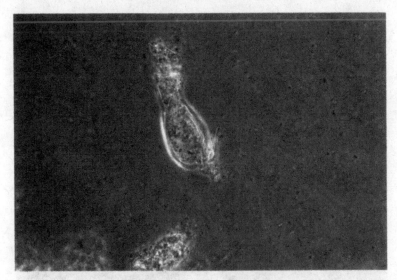

FIGURE 19.7   Dispersed rotifer.

Specific Oxygen Uptake Rate (SOUR)

Because toxicity affects cellular structure and cellular activity, a decrease in the number of active bacteria in the activated sludge process occurs. With a smaller number of active bacteria, less oxygen is consumed as less BOD is degraded. Therefore, a higher dissolved concentration is maintained in the activated sludge process and a decreased specific oxygen uptake rate (SOUR) occurs (Tables 19.19 and 19.20). SOUR (mg/hr/g VSS) is determined by the following calculation:

DO uptake rate (mg/liter)/minute × 60 (minutes/hour) × 1000 mg/g/VSS (mg/liter)

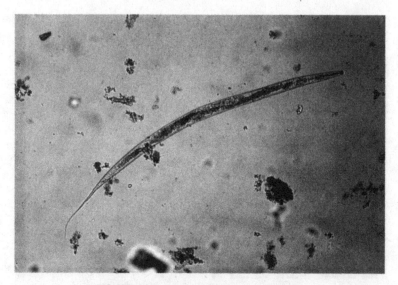

FIGURE 19.8 Healthy free-living nematode.

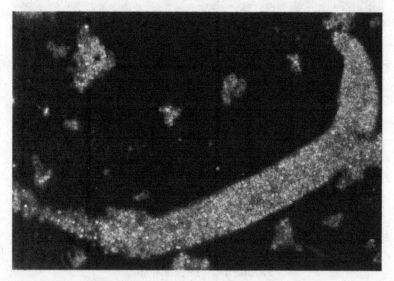

FIGURE 19.9 Dispersed free-living nematode.

TABLE 19.19   Significance of SOUR Values

Values (mg/hr/g VSS)	Rate of Oxygen Consumption	Significance
>20	High	Not enough solids for the BOD loading
12–20	Normal	Good BOD removal and sludge settling
<12	Low	Too many solids or presence of toxicity

TABLE 19.20 Typical Ranges of Specific Oxygen Uptake Rates (SOUR) for Various Modifications of the Activated Sludge Process at Aeration Tank Effluent[a]

Process Modification	SOUR Range (mg/hr/g VSS)
Conventional	8–20
Step aeration	8–20
Extended aeration	3–12
Contact stabilization	5–15
	15–30

[a] Values should be relatively consistent—that is, +2–3 mg/hr/g VSS. SOURs should be taken from the same aeration tank at approximately the same time each day (nearly identical loading conditions on a daily basis) for troubleshooting purposes including the identification of toxicity.

TABLE 19.21 Examples of the Ranges in Concentrations (mg/liter) of Heavy Metals that Cause Toxicity

Metal	cBOD Removal	nBOD Removal
Copper	0.1–1	0.005–0.5
Nickel	1–2.5	0.25–5
Zinc	0.3–20	0.01–1

Nitrification

Degradation of nBOD occurs through two biologically mediated reactions. First, *Nitrosomonas* and *Nitrosospira* oxidize ionized ammonia to nitrite (Equation 19.4). Second, *Nitrobacter* and *Nitrospira* oxidizes nitrite ions to nitrate ions (Equation 19.5). Because *Nitrobacter* and *Nitrospira* are not as tolerant of toxicity as *Nitrosomonas* and *Nitrosospira*, the nitrite produced by *Nitrosomonas* and *Nitrosospira* accumulates due to the inability of *Nitrobacter* and *Nitrospira* to oxidize nitrite as rapidly as *Nitrosomonas* and *Nitrosospira* oxidizes ionized ammonia. This inability due to toxicity is a form of incomplete nitrification.

$$NH_4^+ + 1.5O_2 \xrightarrow{\textit{Nitrosomonas and Nitrosospira}} NO_2^- + 2H^+ + H_2O \qquad (19.4)$$

$$NO_2^- + 0.5O_2 \xrightarrow{\textit{Nitrobacter and Nitrospira}} NO_3^- \qquad (19.5)$$

WHEN DOES TOXICITY OCCUR?

The concentration at which various toxic wastes adversely affect biological processes cannot be stated accurately (Table 19.21). Toxicity is influenced by several operational factors including:

- Form of the toxic waste
- Mean cell residence time (MCRT)
- Mixed liquor volatile suspended solids concentration (MLVSS)

- Mode of operation
- pH of the biological reactor (e.g., aeration tank or anaerobic digester)
- Soluble heavy metals
- Species of dominant bacteria
- Toxic mass-to-biomass ratio

Form of the Toxic Waste

In order for toxic wastes to exert toxicity, the wastes must enter bacterial cells and cause damage. Toxic wastes only can enter bacterial cells in the soluble form. Also, whether the toxic waste is ionized or non-ionized also affects it toxicity. Soluble forms of the toxic wastes are critically important with respect to heavy metal toxicity, while non-ionization of toxic wastes is important with respect to toxicity from ammonia, hydrogen cyanide, hydrogen sulfide, and chlorine.

Mean Cell Residence Time and Mixed Liquor Volatile Suspended Solids

Activated sludge processes that operate at high MCRT as compared to those that operate at low MCRT have a larger number of bacteria (MLVSS) and a larger quantity of inert solids. Although many bacteria and inert solids react with toxic wastes, activated sludge processes that operate at high MCRT will have more active bacteria that survive a toxic occurrence as compared to those that operate at low MCRT. The larger the number of surviving bacteria, the greater the opportunity for the activated sludge processes to maintain effective treatment of wastes or recover rapidly from the toxic occurrence.

Mode of Operation

Activated sludge processes can be operated in different modes including complete mix and plug-flow (Figure 19.10). Each mode of operation impacts differently the significant components in a waste stream. Complete mix mode of operation dilutes nutrients, substrate (food), and toxic wastes into each aeration tank. The dilution of toxic wastes may help to prevent toxicity in the activated sludge process. Plug-flow mode of operation establishes nutrient, substrate, and toxic waste gradients from the first aeration tank to the last aeration tank. The toxic waste gradient may promote toxicity in the upstream aeration tanks of the activated sludge process.

pH of the Biological Reactor

Changes in pH of the biological reactor (aeration tank or anaerobic digester) change the chemical form of several wastes and can make them more or less toxic or nontoxic. For example, the toxic effects of ammonia, chlorine, cyanide, and hydrogen sulfide are influenced by the pH of the biological treatment unit. All of these compounds are more toxic in their undissociated (non-ionized) molecular state (Table 19.22).

The degree of ionization of wastes in biological reactors is influenced by pH. With increasing pH, ammonia (NH_3) is produced in large quantities (Equation 19.6). With

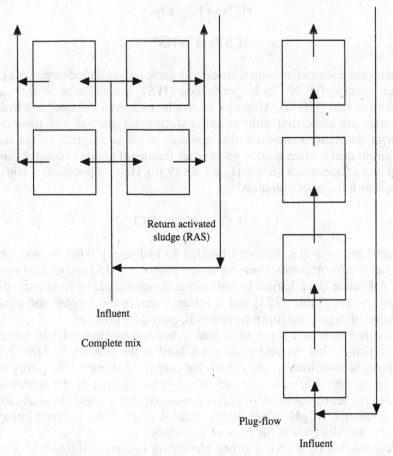

Return activated
sludge (RAS)

Influent

Complete mix

Plug-flow

Influent

FIGURE 19.10 Complete mix and plug-flow modes of operation.

TABLE 19.22 Non-ionized and Ionized Molecular States of Some Toxic Wastes

Non-ionized Molecular State of Toxic Waste		Ionized Molecular State of Toxic Waste	
Name	Formula	Name	Formula
Ammonia	NH_3	Ammonium ion	NH_4^+
Hypochlorous acid	HOCl	Hypochlorous ion	OCl^-
Hydrogen cyanide	HCN	Cyanide	CN^-
Hydrogen sulfide	H_2S	Sulfide	HS^-

decreasing pH, hypochlorous acid (HOCl), hydrogen cyanide (HCN), and hydrogen sulfide (H_2S) are produced in large quantities (Equations 19.7, 19.8, and 19.9).

$$NH_3 + H^+ \leftrightarrow NH_4^+ \tag{19.6}$$

$$HOCl \leftrightarrow H^+ + OCl^- \tag{19.7}$$

$$HCN \leftrightarrow H^+ + CN^- \tag{19.8}$$

$$H_2S \leftrightarrow H^+ + HS^- \tag{19.9}$$

Examples of common, non-metal toxic, inorganic wastes include ammonia (NH_3), hydrogen cyanide (HCN), hydrogen sulfide (H_2S), and chlorine in the form of hypochlorous acid (HOCl). Although ammonia, hydrogen cyanide, and hydrogen sulfide often are associated with industrial wastewater and the anaerobic decomposition of domestic wastewater that contains proteins, chlorine in the form of hypochlorous acid is often used in wastewater treatment plants to control malodors and undesired filamentous growth (Equation 19.10). Here, hypochlorous acid comes in contact with the active biomass.

$$Cl_2 + H_2O \rightarrow HCl + HOCl \tag{19.10}$$

Reduced nitrogen (i.e., nitrogen bonded to hydrogen) exists in two forms in wastewater treatment plants. These forms are ammonia (NH_3) and ionized ammonia (NH_4^+). Ammonia is discharged to wastewater treatment plants from domestic and industrial sources (Table 19.23) and is released during the aerobic and anaerobic degradation of organic-nitrogen compounds, especially proteins.

The ionized ammonia is not toxic and serves two positive roles in wastewater treatment plants. First, ionized ammonia is used as the primary bacterial nutrient for nitrogen. Second, ionized ammonia is the energy substrate for the nitrifying bacteria in the genera *Nitrosomonas* and *Nitrosospira*. The pH of the wastewater or sludge determines the quantity of reduced nitrogen that is presence as ammonia or ionized ammonia. At pH values greater than 9, most of the reduced nitrogen in wastewater and sludge is in the form of ammonia.

The cyanide ion (CN^-) has a strong affinity for many metal ions and is used in industrial applications, especially metal cleaning, electroplating, and mineral processing. Cyanide is dissociated or ionized, while hydrogen cyanide (HCN) is undis-

TABLE 19.23 Significant Industrial Sources of Nitrogenous Wastes

Automotive facilities
Chemical manufacturing
Coal gasification
Fertilizer manufacturing
Food processing facilities
Landfills
Livestock maintenance
Meat processing
Ordnance sites
Petrochemical
Pharmaceutical manufacturing
Primary metal industries
Refineries
Steel manufacturing
Tanneries

sociated. The undissociated form is highly toxic. In water, hydrogen cyanide exists as a weak acid and dissociates. With decreasing pH, less dissociation of hydrogen cyanide occurs, and more of the toxic hydrogen cyanide exists.

Dissociation of hydrogen cyanide results in the production of cyanide. Cyanide has a strong affinity for metals, especially iron, and quickly forms ferrocyanide ($Fe(CN)_6^{4-}$) in the presence of iron. Ferrocyanide as well as other cyanide-containing (cyano-) compounds are toxic.

Hydrogen sulfide (H_2S) is produced under septic conditions through the anaerobic degradation of organic compounds that contain sulfur such as proteins. It also is produced when sulfate (SO_4^{2-}) is used by sulfate-reducing bacteria when they degrade organic compounds. Hydrogen sulfide also may be found in wastewater from chemical plants, paper mills, tanneries, and textile mills. The presence of hydrogen sulfide in wastewater is easily detected by its characteristic rotten-egg odor.

Reduced sulfur (i.e., sulfur bonded to hydrogen) exists in two forms. These forms are hydrogen sulfide (H_2S) and sulfide (HS^-). In the undissoicated form, hydrogen sulfide is highly toxic. In water, hydrogen sulfide is a weak acid and dissociates to form the sulfide ion (HS^-). The pH of the wastewater or sludge determines the quantity of H_2S and HS^-. At pH values less than 6, nearly all of the reduced sulfur is present as hydrogen sulfide. Hydrogen sulfide is sparingly soluble in water. Therefore, it will partition between water and gas phases.

Heavy Metals

Metals that have a significant detrimental impact upon biological treatment units are referred to as "heavy" metals. These metals cause the following undesired consequences when present in excessive quantities:

- Toxicity to organotrophic bacteria in aerobic treatment reactors
- Toxicity to organotrophic bacteria in anaerobic treatment reactors
- Toxicity to nitrifying bacteria in aerobic treatment reactors
- Interruption of floc formation
- Permit violations
- Accumulation of metals in sludges
- Increased operational costs

Metals such as cadmium (Cd), copper (Cu), lead (Pb), mercury (Hg), nickel (Ni), silver (Ag), and zinc (Zn) are well-known, toxic heavy metals. These metals can occur in a variety of wastes and cause acute or chronic toxicity. In order for heavy metals to exert their toxic impact, the metals must be soluble and must be present as a free ion (e.g., Cu^{2+}) or a metal oxide (e.g., CuO). As free ions or metal oxides the metals can be absorbed by bacteria and then exert their toxic effects (Figure 19.11).

The adsorption or uptake of heavy metals is very rapid. The rapid uptake is considered to occur over 30 minutes in two phases (Figure 19.12). During the first phase of metal uptake (approximately 10 minutes), most of the metals in a biological process are removed from the bulk solution through adsorption to bacterial fibrils. During the second phase of metal uptake (approximately 20 minutes), most of the

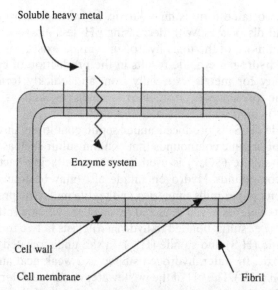

Soluble heavy metal

Enzyme system

Cell wall

Cell membrane

Fibril

FIGURE 19.11 *Heavy metal uptake by a bacterial cell. Heavy metals are transported into a bacterial cell through their adsorption to bacterial fibrils and then absorption into the cytoplasm.*

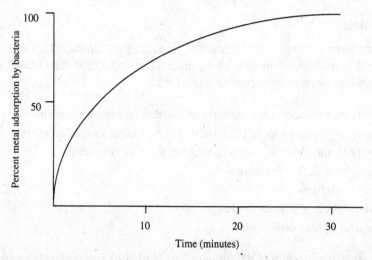

FIGURE 19.12 *Two-phase adsorption of heavy metals by bacteria in the activated sludge process. Most heavy metals are removed efficiently in the activated sludge process over a two-phase adsorption period. During the first phase of 10 minutes, over 70% of the influent metals to the activated sludge process are adsorbed to bacterial fibrils or inert solids. During the second phase of 20, many of the remaining heavy metal are adsorbed to bacterial fibrils or inert solids.*

remaining metals in the bulk solution are adsorbed to bacterial fibrils. The uptake of heavy metals in biological processes is very efficient. In the activated sludge process, uptake or removal efficiency for many metals often approaches 99%. The rapid and efficient removal of heavy metals is indicative of the susceptibility of biological processes to heavy metal toxicity.

There are approximately 55 heavy metals. Several of these metals are probably the most studied toxic wastes with respect to their impact upon biological wastewater treatment processes. There are two definitions that are used to describe heavy metals. First, the metals can be precipitated by hydrogen sulfide in an acidic solution. Second, the metals are members of the metals and metalloids in the groups IB, IIB, IIIB, IVB, VB, VIB, and VIII of the Periodic Table of Chemical Elements.

Although some metals such as nickel are oxidized by bacterial cells and are incorporated into bacterial enzyme systems to enhance enzymatic activity, excess soluble metals may cause toxicity. When metals are incorporated into enzyme systems, the metals are referred to as additives or activators.

The major source of toxic metals is industrial wastewater. For example, metal plating processes can discharge aluminum (Al), cadmium (Cd), chromium (Cr), copper (Cu), iron (Fe), nickel (Ni), silver (Ag), and zinc (Zn) in addition to cyanide. Steel mill wastewater can contains aluminum, iron, and magnesium (Mg).

Metals in the influent of a wastewater treatment plant are removed or partitioned in different treatment tanks. Approximately 20% of the influent metals are removed in sedimentation tanks or primary clarifiers (Figure 19.13). Metal removal here is physical. Settling of metals occurs in primary clarifiers, if the metals are in the insoluble or precipitated form or are adsorbed to solids that settle.

Metals pass through the primary clarifiers and enter the activated sludge process because the metals are in solution as free ionic state species (e.g., Cu^{2+}) or metal oxides, or they are bonded to chelating agents or ligands (Table 19.24). Ligands are soluble and may be simple or complex in structure and may be inorganic or organic in composition. Regardless of the structure or chemical composition of ligands, ligands hold metals in solution. If the metals stay in solution, the metals cannot cause toxicity but do pass through the wastewater treatment plant.

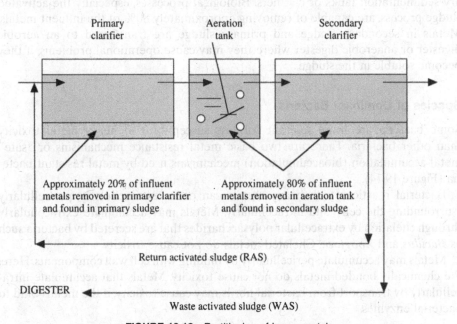

FIGURE 19.13 Partitioning of heavy metals.

TABLE 19.24 Ligands Commonly Found in Wastewater Treatment Plants

Ligand Strength	Examples
Weak binding ligands	Ionized ammonia (NH_4^+)
	Carbonate (CO_3^{2-})
	Chloride (Cl^-)
	Hydroxide (OH^-)
	Phosphate (PO_4^{3-})
	Sulfate (SO_4^{2-})
Strong binding ligands	Fulvic acid (soil component)
	Humic acid (soil component)

Metals may remain bonded to ligands or may be removed from the ligands by solids in the biological process, if solids have a stronger affinity for the metals than the ligands have. If the metals remain bonded to the ligands, toxicity is prevented, but the metals pass through the biological process to the receiving body of water. If metals from the ligands, free ionic state species of metals, or metal oxides are adsorbed by solids in the biological process, toxicity may occur.

For metal toxicity to occur in a biological treatment unit, metals must be adsorbed by the bacteria. The adsorbed metals then must be absorbed and must attack enzymes within the bacterial cells. Insoluble metals, metals bonded to ligands, and soluble metals that are adsorbed by nonbacterial or inert solids such as particulate materials do not cause toxicity.

Metals removed in the activated sludge process through their adsorption to bacteria and inert solids are concentrated in the settled solids or sludge in the secondary sedimentation tanks or clarifiers. Biological processes, especially the activated sludge process, are capable of removing approximately 80% of the influent metals. Metals in secondary sludge and primary sludge are transferred to an aerobic digester or anaerobic digester where they may cause operational problems, if they become soluble in the sludge.

Species of Dominant Bacteria

Some bacteria are more tolerant (or less susceptible) to heavy metal toxicity than other bacteria. There are two basic metal resistance mechanisms or "safe" metal accumulation (bioaccumulation) mechanisms used by metal resistant bacteria (Figure 19.14).

Bacterial reactions with heavy metals can occur extracellularly, pericellularly (surrounding the cell), and intracellularly. Metals may accumulate extracellularly through chelation by extracellular polysaccharides that are secreted by bacteria such as *Bacillus* and *Zoogloea*. Chelated metals do not cause toxicity.

Metals may accumulate pericellularly by binding with cell wall components. Here, the chemically bonded metals do not cause toxicity. Metals that accumulate intracellularly by transport from bacterial fibrils may cause toxicity, if the metals bind to bacterial enzymes.

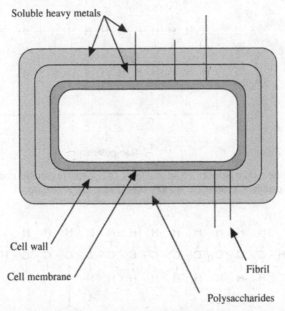

FIGURE 19.14 *Bioaccumulation of heavy metals. Bacterial can remove or accumulate heavy metals "safely" through the adsorption of heavy metals to the polysaccharides surrounding the bacteria or cell wall. The adsorption of heavy metals to bacterial fibrils is "unsafe" due to the absorption to the cytoplasm and "attack" upon enzymes.*

Toxic Mass-to-Biomass Ratio

Perhaps the most important operational factor affecting toxicity in a biological treatment unit is the toxic mass-to-biomass ratio. The lower this ratio, the better the process is able to tolerate or treat a toxic event and the greater the number of bacteria that will survive a toxic event. A low toxic mass-to-biomass ratio can be achieved by reducing the concentration of the toxic waste entering the biological reactor and/or increasing the number of bacteria (solids) in the biological reactor.

SURFACTANTS

Surfactants are the key ingredient of detergents. Surfactants are the surface-active agents, which make water "wetter"—that is, a better cleaning agent. Surfactants concentrate at the interfaces of water with gases, solids (dirt), and immiscible liquids (oils).

The ability of surfactants to concentrate at an interface is due to their molecular structure (Figures 19.15 and 19.16). Surfactants contain a polar or ionic group (head) with a strong affinity for water and a nonpolar or non-ionic group (tail) with an aversion to water. The tail is a hydrocarbon group.

Although surfactants contribute to operational problems, other ingredients in detergents also contribute to operational problems. Most commercial solid detergents contain 10–30% surfactant. An additional critical component of a detergent

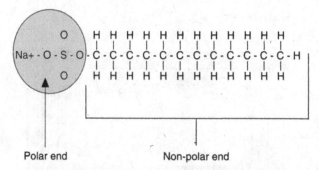

Polar end   Non-polar end

FIGURE 19.15 *Lauryl sulfate (sodium dodecylsulfate). Lauryl sulfate (sodium dodecylsulfate) is used in a large variety of cleaners, cosmetics, laundry detergents, and shampoos.*

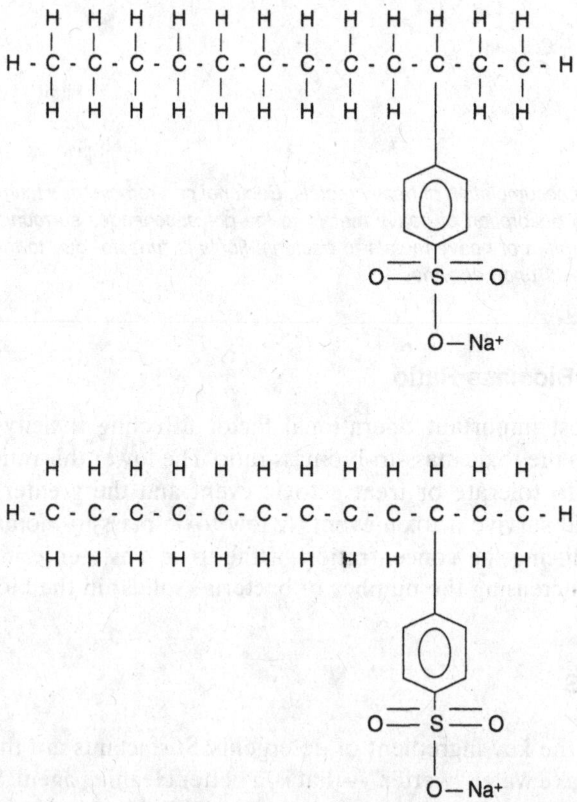

FIGURE 19.16 *Isomers of LAS (linear sulfonate). LAS has a benzene ring that may be attached anywhere on the carbon chain between the two terminal carbon atoms. LAS is used in many laundry detergents.*

is the "builder." An example of a builder is polyphosphate. The builder binds to hardness ions such as calcium in the water. The builder makes the detergent solution alkaline and improves the action of the surfactant.

Other ingredients in detergents may include alkalis, anticorrosive compounds, bleaches, brighteners, dyes, fabric softeners, foam stabilizers, fragrances, ion exchangers, soil-suspending compounds, and enzymes to degrade lipids (lipase) and starches (cellulases). Several undesired effects of slowly degrading detergents or slug discharges of detergents include

- Deflocculation of the biomass
- Deflocculation of colloids
- Emulsification of fats, oils, and grease
- Flotation of solids
- Foam production (decreased surface tension of wastewater)
- Toxicity

Toxicity caused by surfactants is influenced by several factors including the molecular structure of the surfactant, water hardness, temperature, and dissolved oxygen. The most important factors are the molecular structure of the surfactant and dissolved oxygen concentration.

Surfactant toxicity is probably due to the damage that surfactants cause to cellular proteins and the cell membrane. Even when surfactants present no toxicity concern, low levels of surfactants may increase the uptake of other wastes.

With respect to molecular structure, surfactants can be broadly placed into three groups, anionic, non-ionic, and cationic. Anionic surfactants are the most widely used and the most commonly discharged surfactants to wastewater treatment processes. A commonly used group of surfactants are the sulfonates.

Non-ionic surfactants have attracted less attention than anionic surfactants. Generally, non-ionic surfactants as well as anionic surfactants tend to be more toxic at lower concentrations than cationic surfactants. Cationic surfactants are used mostly as medical and laboratory disinfectants.

Surfactants differ from many toxic wastes. They are not uniformly distributed in water. They concentrate at surfaces. An additional difference is that one surfactant may exist in a large number of isomers or molecular structures. These isomers are responsible for the variation in reported results of surfactant toxicity. Also, loss of surfactant toxicity during testing may occur as surfactants are degraded in all but short testing periods.

The molecular structure of a surfactant, especially anionic surfactants, greatly influences toxicity. LAS (1-benzenesulfonate), for example, has a benzene ring that may be attached at any point on the alkyl (carbon) chain except the terminal or end carbons. A "hard" surfactant such as ABS (alkylbenzenesulfonate) is branched and the benzene ring is attached at an end or terminal carbon unit on the alkyl chain. A hard surfactant is a persistent or slowly degradable surfactant. Significant impacts of hard surfactants are undesired foaming and destruction of bacteria, protozoan, and metazoa.

Water hardness also affects surfactant toxicity. The toxicity of an anionic surfactant may increase or decrease with increasing water hardness. The toxicity of an non-ionic surfactant usually is not affected by charges in water hardness.

The toxicity of surfactants generally increases with increasing wastewater temperature. The increase in toxicity may be due to the increase in rate at which the surfactant attacks the cell and the decrease in ability of the wastewater to hold dissolved oxygen with increasing temperature. Low concentrations of anionic surfactants and non-ionic surfactants that are not toxic during cold wastewater temperatures may be toxic during warm wastewater temperatures.

Dissolved oxygen concentration affects the toxicity of surfactants as well as other toxic wastes. With increasing dissolved oxygen concentration, surfactants exert decreasing toxicity.

Salinity consists of dissolved sodium (Na^+), potassium (K^+), and magnesium (Mg^{2+}). Salinity influences the osmoregulatory ability of a cell—that is, the ability of the cell to import materials and export wastes. During stress conditions such as the presence surfactants, changes in salinity may increase or decrease the impact of surfactant toxicity.

Total suspended solids (TSS) also influence surfactant toxicity. Because some suspended solids such as kaolin clay adsorb surfactants, increasing TSS concentration may reduce surfactant toxicity. The adsorbed surfactants are removed from the waste stream when the solids settle in the secondary clarifiers.

CONTROLLING TOXICITY

There are several operational measures that can be used to minimize or prevent toxicity in a biological treatment process. These measures include the following:

- Identify potential toxic dischargers.
- Identify potential toxic wastes.
- Monitor and regulate the discharge of potential toxic wastes.
- Develop simplistic and reliable indicators of toxicity in the treatment process.
- Regulate solids inventory in reaction tanks.
- Use granular activated carbon (GAC).
- Use coagulants and polymers.

Potential toxic dischargers include industrial and commercial establishments. These establishments can be identified by the quantity and quality of wastes discharged to the sanitary sewer, the time of day of discharge, and whether a discharge produced an unacceptable condition such as a high pH or low pH within the sanitary sewer or treatment process. Identification of potential toxic dischargers and toxic wastes should incorporate the periodic review of material safety data sheets (MSDS). The review should include MSDS of all cleaning agents that are used at industrial and commercial establishments including hospital and school cafeterias and hospital laundry facilities.

The discharge of potential toxic wastes should be closely monitored and regulated to prevent toxicity in the treatment process. The discharge should be regulated or equalized to ensure a constant low concentration of the toxic waste enters the treatment process over designated time periods. Careful regulation of the discharge of a toxic waste can acclimate the treatment process to accept higher concentrations of the toxic waste over time.

Acclimation of the treatment process is slow and develops a biomass that can survive and treat wastes efficiently in a new environment or condition such as gradually increasing concentrations of a toxic waste. Acclimation should be quickly terminated if signs of toxicity occur. Shock loads of toxic wastes are unacceptable and should be avoided. If a shock load should occur, the load may be diverted to a temporary storage facility and then slowly discharged to the treatment process.

Simplistic and reliable indicators of toxicity should be developed and used on a routine or as needed basis in order to quickly identify toxicity and initiate appropriate process control measures. These indicators may be biological, chemical, or physical and should be applicable for the activated sludge process (Table 19.14) and the anaerobic digester (Table 19.15).

Microscopic indicators of toxicity can be mimicked for training purposes. For example, healthy mixed liquor should be observed under the microscope and then the mixed liquor can be "spiked" with a suspect toxic waste such as a heavy metal or surfactant that may enter the treatment process as identified by an MSDS. After the healthy mixed liquor has been spiked with the toxic waste, the mixed liquor and toxic waste should be allowed to react during an aeration period and a sample of spiked mixed liquor can then be examined microscopically. The healthy mixed liquor and spiked mixed liquor should be examined for differences in (a) quantity of dispersed growth and particulate material in the bulk solution, (b) shapes, sizes, and strength of floc particles, and (c) activity and structure of protozoa and metazoa.

Solids inventory in the activated sludge process can be regulated to either degrade the toxic wastes entering the treatment process or minimize or prevent toxicity. Solids inventory can be regulated through two techniques.

First, aeration tanks that typically are not used may be used to address toxicity concerns. Uniform solids concentration should be maintained in all aeration tanks. Aeration tanks (reserve solids or bacteria) that are not needed for daily treatment of wastewater should be placed "off-line" by closing the influent gate to these tanks. The off-line aeration tanks should be aerated and may be rotated daily (Figure 19.17). When a toxic waste enters the activated sludge process the off-line tanks may be placed "on-line" quickly by opening the influent gates. By placing reserve solids or bacteria on-line, the following conditions are established:

- The toxic mass-to-biomass ratio is lowered.
- More bacteria are available to degrade the toxic waste.
- An increase in aeration tank capacity (volume) that is produced by placing more tanks on-line provides for increased dilution of the toxic waste.

Second, reserve solids may be stored in the contact-stabilization mode of operation (Figure 19.18). Solids in the stabilization tanks are maintained at a higher concentration than solids in the contact tanks. Solids in the stabilization tanks are protected from a direct "hit" of toxic wastes.

Although off-line aeration tanks may not be needed to overcome toxicity for relatively long periods of time, the off-line aeration tanks provide for the following benefits:

- Increased volatile solids destruction
- Use of nitrate (NO_3^-) produced through nitrification for anoxic periods of 1–2 hours to destroy undesired filamentous organism growth
- Use of nitrate ions to strengthen floc particles

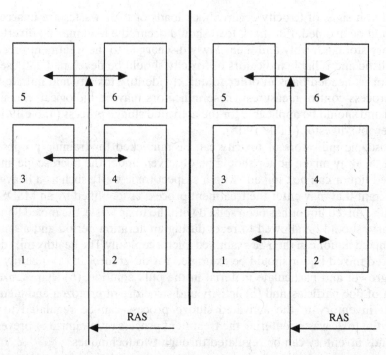

FIGURE 19.17   *Rotating aeration tanks. Aeration tanks 1 and 2 are taken off-line during a fixed period of time (left), and then aeration tanks 1 and 2 are placed back on-line and aeration tanks 3 and 4 are taken off-line for a period of time (right).*

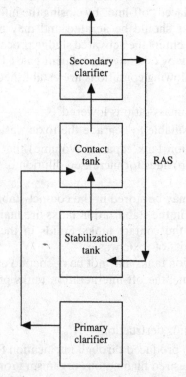

FIGURE 19.18   *Contact stabilization mode of operation.*

• Cycling of nitrification and denitrification (anoxic) periods to reduce total nitrogen discharged from the treatment process

Granular activated carbon (GAC) can be used to overcome toxicity. GAC improves floc formation and removes many toxic organic compounds and some toxic inorganic compounds before they come in contact with large numbers of bacteria. GAC also helps to develop a dense or highly populated biomass that lowers the toxic mass-to-biomass ration. Coagulants and polymers also improve floc formation and remove some toxic wastes.

RECOVERY FROM TOXICITY

There are two operational measures that can be used to recover from toxicity. These measures include reseeding the treatment process and reacclimating the new biomass. Reseeding an activated sludge process can be achieved by introducing activated sludge from another treatment process. The introduced sludge may be from an activated sludge process or aerobic digester. An anaerobic digester may be reseeded with sludge from another anaerobic digester or with fresh cow manure. Reseeding also may be achieved with bioaugmentation products.

For anaerobic digesters, reseeding can be achieved with the addition of primary clarifier sludge from another treatment process or fresh cow manure. Cow manure is rich in methane-forming bacteria. Approximately 5 gallons of cow manure should be added for each 100,000 gallons of digester sludge.

Bioaugmentation is the addition of commercially prepared bacterial cultures to biological treatment processes. The bacterial cultures can be used to enhance the performance of an indigenous biomass with species of bacteria that have specific degradative capabilities or tolerances including degradation of specific wastes or fluctuations in concentrations of toxic wastes.

Bioaugmentation products are available as dry products and liquid products (Table 19.25). Although bioaugmentation products typically are described as containing bacteria, some products contain fungi. The addition of fungi to biological treatment processes usually is performed to enhance the degradation of cellulose, dyes, and lignin. Examples of fungi placed in bioaugmentation products include *Aspergillus*, *Ceriporiopsis*, and *Trichoderma*.

Bioaugmentation products can be used to treat a variety of toxic wastes. Examples of industrial wastes that contain toxic wastes that can be treated with bioaugmentation products include citrus fruit processing (press liquor), organic chemicals (synthetic), pulp and paper (black liquor), and petrochemical. Examples of specific, toxic organic compounds that can be treated with bioaugmentation products include

TABLE 19.25 Basic Components of Bioaugmentation Products

Packaging	Bacteria	Additives
Dry	Lyophilized cells (freeze dried)	Macronutrients, micronutrients, enzymes, odor binders for amino groups and thiol groups
Liquid	Preserved and unpreserved cells	Surfactants, enzymes, nitrate, and odor binders for amino groups and thiol groups

acetone, acrylic acid, ammonia, nitrite, dimethyformamide, furfural, phenols, phenolic compounds, and methyl ethylamine.

INTERACTIONS BETWEEN WASTES

Within biological treatment units, bacteria are exposed to many wastes simultaneously. Interactions between these wastes can either increase (synergistic effect and potentiation) or decrease (antagonistic effect) the toxicity of the wastes. The mode of interaction between different wastes can be due to the molecular structure of the wastes or changes in the metabolic processes in the organisms.

A synergistic effect or response occurs when the combined effect of two wastes is greater than the sum of the individual wastes. A synergistic response can be mimicked by potentiation. Potentiation occurs when one of the wastes to which the organisms are exposed is nontoxic alone, but in the presence of a toxic waste the nontoxic waste enhances the effect of the toxic waste. For example, nontoxic isopropanol ($CH_3CH_2CH_2OH$), when present with toxic carbon tetrachloride (CCl_4), enhances the toxic effect of carbon tetrachloride.

An antagonistic effect or response occurs when one waste interferes or reduces the toxic effect of another waste. This occurs when the two wastes interact at the same site of the organism. Different types of antagonistic responses occur. Commonly occurring antagonist responses include chemical antagonism and competitive antagonism. Chemical antagonism occurs when one waste (an antagonist) inactivates a toxic waste (antagonist) through a chemical reaction. For example, selenium (Se) binds with toxic mercury (Hg) and prevents the binding of mercury with the thiol group (-SH) on cellular proteins. The action of selenium prevents protein damage. Competitive antagonism occurs when one waste displaces a toxic waste from its cellular site of attack.

MICROBIAL BIOASSAYS

A bioassay is an interpretation of the response of organisms to chemicals in their environment. Often microbial bioassays are used to protect an activated sludge process (mixed liquor biota) from toxic industrial wastes. Microbial bioassays can be used to identify influent toxicity and determine the ability of the activated sludge process to tolerate or reduce toxicity.

· Microbial bioassays usually measure bioluminescence, respiration rate, or substrate utilization rate. A reduction in bioluminescence, rate of respiration, or substrate utilization usually is indicative of toxicity. Microtox® testing and respiration rate (respirometry) are the commonly used microbial bioassays.

The Microtox® test (Microtox® Toxicity Test System) measures the quantity of light emitted by the bioluminescent (luminescent) bacterium *Vibrio fischeri* (*Photobacterium phosphoreum*). This bacterium uses some of its respiratory energy in a metabolic pathway that converts chemical energy into visible light. Any change in cellular metabolism or cellular structure due to the addition of a toxic waste results in depressed cellular respiration and a decrease in the intensity of visible light. The more toxic the waste, the greater the percent light loss from the test suspension of luminescent bacteria.

The respiration rate measures the rate of oxidation of a readily degradable substrate such as glucose to carbon dioxide in two samples. These samples include a return activated sludge (RAS) sample diluted with tap water to mixed liquor suspended solids (MLSS) concentration in the activated sludge process and a RAS sample diluted with influent that contains the suspect toxic waste. Each RAS sample is fed the same readily degradable substrate. If the respiration rate of the RAS-influent sample is less than the RAS-tap water sample, the influent is considered to be toxic.

The substrate utilization rate (Delta COD Test) determines the impact of a suspect toxic waste in the influent upon the ability of the biomass to degrade substrate. The substrate utilization rate uses a series of tests to compare the rate that the biomass uses substrate (COD) to the rate that is obtained when the biomass is fed plant influent that contains the suspect toxic waste.

Part VII

Foam and Malodor Production

20

Microbial Foam

The production of microbial (bacterial) foam in the activated sludge process and anaerobic digester usually is the result of undesired bacterial activity. Each foam in the activated sludge process has a characteristic texture and color (Table 20.1). The texture and color of foam may be different from its description due to the sludge age of the activated sludge process and the presence of another foam. Foam becomes more viscous and darker in color with increasing sludge age and more billowy and lighter in color with decreasing sludge age. The presence of another foam results in the blending of textures and colors. Anaerobic digester foam (Table 20.2), regardless of its contributing bacterial activity, is black due to the septic nature of the digester sludge. The production and accumulation of any microbial foam at a biological treatment unit may contribute to increased operational costs and permit violations.

Foam consists of entrapped air or gas bubbles beneath a thin layer of solids or biological secretions. Gases entrapped in foam consist of those released during the aerobic and anaerobic degradation of soluble cBOD. Major gases entrapped in foam include carbon dioxide and molecular nitrogen (N_2).

ACTIVATED SLUDGE FOAM

Microbial foam is produced in the aeration tank (Figure 20.1) and is discharged from the aeration tank to other treatment units such as the secondary clarifier, thickener, and dewatering facilities. As foam is discharged from the aeration tank to other treatment units, the entrapped air and gas bubbles escape, and the foam collapses. The collapsed foam often is referred to as scum (Figure 20.2).

Wastewater Bacteria, by Michael H. Gerardi
Copyright © 2006 John Wiley & Sons, Inc.

TABLE 20.1   Microbial Foam Production in the Activated Sludge Process

Contributing Biological Activity	Texture and Color
Erratic sludge wasting rates	Concentric circles of viscous dark brown and viscous light brown foam
Foam-producing filamentous bacteria	Viscous chocolate brown
Nutrient deficiency, old sludge	Greasy gray
Nutrient deficiency, young sludge	Billowy white
Sludge aging	Billowy white to crisp white to crisp brown to viscous dark brown
Slug discharge of soluble cBOD	Billowy white
Zoogloeal growth (viscous floc)	Billowy white

TABLE 20.2   Microbial Foam Production in the Anaerobic Digester

Increase in alkalinity
Increase in carbon dioxide
Increase in fatty acids, including start-up condition
Particulate surfactants
Polymers
Temperature fluctuations

FIGURE 20.1   *Foam on an aeration tank.*

Erratic Wasting Rates

Erratic wasting rates of solids from the activated sludge process results in the production of "pockets" of young bacterial growth and old bacterial growth. Because young bacterial growth does not produce and accumulate large quantities of oils as compared to old bacterial growth, air and gas bubbles captured by young and old

FIGURE 20.2   *Collapsed foam or "scum" on the surface of a secondary clarifier.*

floc particles result in the production of a different foam for each pocket of growth. Viscous light brown foam is produced for young bacterial growth and viscous dark brown foam is produced for old bacterial growth. Concentric circles of foam can be observed on the surface of the aeration tank when the aeration and mixing are terminated (Figure 20.3). A steady or consistent wasting rate over a relatively long period of time can help to correct this foam condition.

Foam-Producing Filamentous Bacteria

The two most commonly occurring foam-producing filamentous bacteria in the activated sludge process in North America are *Microthrix parvicella* (Figure 20.4) and Nocardioforms (Figure 20.5). Foam production by *Microthrix parvicella* and Nocardioforms occurs through different biological processes. *Microthrix parvicella* is hydrophobic and captures air and gas bubbles that result in the production of foam. Foam production from Nocardioforms is the result of (1) the secretion of lipids by living cells that coat the floc particles and capture air and gas bubbles and (2) the release of biosurfactants (e.g., ionized ammonia) that lower the surface tension of the activated sludge.

The undesired growth of each foam-producing filamentous bacteria can be associated with specific operational conditions (Table 20.3). By monitoring and regulating these conditions, the undesired growth of these organisms and their production of foam can be controlled.

Because foam-producing filamentous bacteria are present in large numbers in the foam, the foam represents a source of "reseeding" of the activated sludge with filamentous bacteria. Therefore, treatment of the foam also should be addressed when attempting to control the growth of foam-producing filamentous bacteria (Table 20.4).

FIGURE 20.3   *Concentric circles of different colors and textures of foam due to erratic wasting rates.*

FIGURE 20.4   Microthrix parvicella.

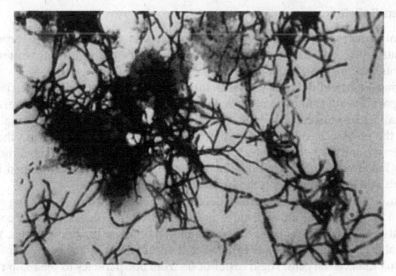

FIGURE 20.5 Nocardioforms.

TABLE 20.3 Operational Conditions Associated with the Undesired Growth of Foam-producing Filamentous Bacteria

	Filamentous Bacterium	
Operational Condition	*Microthrix parvicella*	Nocardioforms
High MCRT (>10 days)	X	
Fats, oils, and grease	X	X
High pH (>8)	X	
Low DO and High MCRT	X	
Low F/M (<0.05)	X	X
Low nitrogen or phosphorus		X
Low pH (<6.5)		X
Readily degradable cBOD		X
Slowly degradable cBOD	X	X
Winter proliferation	X	

TABLE 20.4 Operational Measures Available for the Control of Filamentous Organisms in Foam

Collapse the foam with the application of cationic polymer.
Collapse the foam with the application of final effluent through lawn sprinklers.
Collapse the foam with non-petroleum-based defoaming agent.
Digest the foam with application of bacterial cultures that contain lipase enzymes.
Physically remove the foam.
Treat the foam with 10–15% sodium hypochlorite solution and spray the foam with final effluent approximately 2–3 hours after sodium hypochlorite treatment.

Nutrient Deficiency

Nutrient deficiencies commonly are experienced in activated sludge processes and usually are due to the presence of nutrient-deficient industrial wastewater. Nutrients that are most often deficient are nitrogen and phosphorus.

During a nutrient deficiency, soluble substrate that is absorbed by bacterial cells in floc particles but cannot be degraded. The nondegraded substrate is converted by bacterial cells to insoluble polysaccharides and stored outside the bacterial cells. Often, the polysaccharides are deposited in the floc channels that permit the flow of water, air, and gas bubbles through the floc particle. When these channels become heavily laden with polysaccharides, air and gas bubbles become entrapped in the channels and foam appears on the surface of the aeration tank.

Foam produced during a nutrient deficiency is billowy white at a young sludge age and greasy gray at an old sludge age. The difference in texture and color of nutrient-deficient foam is due to the accumulation of oils in the floc particles. Young bacterial cells produce relatively little oil that accumulates in floc particles as compared to old bacterial cells. The transfer of oil from floc particles to the foam results in the production of greasy gray foam as the sludge age increases.

A nutrient deficiency can be prevented by ensuring the presence of necessary quantities of nutrients in industrial wastewater. Also, nutrients can be added to the activated sludge process. Chemical compounds that release ionized ammonia (NH_4^+) or orthophosphate ($H_2PO_4^-/HPO_4^{2-}$) can be added to the aeration tank influent. Appropriate recycle streams (digester decant, filtrate, and centrate) can be added to the aeration tank influent, if these streams contain adequate nutrients and a relatively low quantity of cBOD.

Sludge Aging

Several types of foam are produced through changes in bacterial activity with increasing or decreasing sludge age. Billowy white foam is produced at a young sludge age when the bacterial population or mixed liquor volatile suspended solids (MLVSS) are relatively small (e.g., <1000 mg/liter). This small population of bacteria lacks adequate enzymatic activity to degrade the surfactants entering the treatment process. The nondegraded surfactants produce billowy white foam. With increasing sludge age, the bacterial population obtains adequate enzymatic activity and degrades the surfactants. With the degradation of the surfactants, the foam becomes crisp white.

As the bacterial population continues to age (or MLVSS increase), large quantities of oils secreted by the bacteria accumulate in floc particles. The color of the oils darkens crisp white foam to crisp brown foam. Finally, as the bacterial population matures (MLVSS peak), large numbers of slow-growing filamentous organisms such as *Microthrix parvicella* and Nocardioforms may proliferate. These organisms contribute to the production of viscous chocolate brown or dark brown foam.

Undesired quantities of foam as a result of young or old sludge ages can be corrected by decreasing or increasing the sludge age (wasting rate) of the treatment process. However, sludge wasting rates should be as uniform as possible over a relatively long time period.

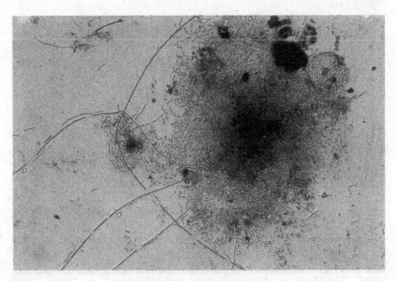

FIGURE 20.6 *Weak and buoyant young growth surrounding firm and dense old growth in floc particle that has experienced a slug discharge of soluble cBOD.*

Slug Discharge of Soluble cBOD

A slug discharge of soluble cBOD is the presence of a quantity of soluble carbonaceous substrate that is two to three times greater than normally received at a biological treatment unit over a period of 2–4 hours. The slug discharge promotes rapid, young bacterial growth (Figure 20.6). This growth contains a copious quantity of insoluble and buoyant polysaccharides that capture air and gas bubbles. The entrapment of air and gas bubbles produces billowy white foam. Foam production from a slug discharge of soluble cBOD can be controlled by preventing the discharge of slug quantities of soluble cBOD from industries.

Zoogloeal Growth (Viscous Floc)

Zoogloeal growth or viscous floc is the rapid and undesired proliferation of floc-forming bacteria such as *Zoogloea ramigera*. Zoogloeal growth may appear in the dendritic or "finger-like" pattern (Figure 20.7) or in the amorphous or globular pattern (Figure 20.8). Zoogloeal growth is associated with the production of large quantities of insoluble gelatinous material (Figure 20.9) that entraps air and gas bubbles. The entrapment of air and gas bubbles results in the production of billowy white foam.

The occurrence of Zoogloeal growth is due to high MCRT, long HRT, nutrient deficiency, organic acids, and significant changes in F/M. Zoogloeal growth and its foam can be controlled by (1) regulating the operational conditions responsible for its proliferation or (2) exposing the growth to anoxic periods.

CAPTURE OF AIR AND GAS BUBBLES BUT NO FOAM

Some bacteria in the activated sludge process are capable of capturing air and gas bubbles, but they do not produce foam. For example, aerobic *Acetobacter* synthe-

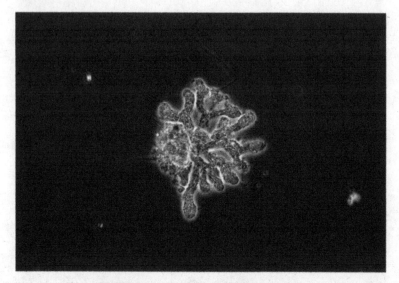

FIGURE 20.7   *"Finger-like" Zoogloeal growth.*

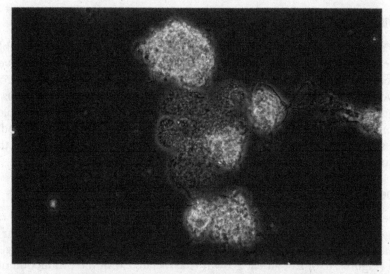

FIGURE 20.8   *Globular or amorphous Zoogloeal growth.*

sizes cellulose. When strands of cellulose build on the cellular surface, they form a mat that captures air and gas bubbles and keeps the *Acetobacter* afloat near the surface of the wastewater where oxygen is most concentrated.

ANAEROBIC DIGESTER FOAM

There are six major types of microbial foam that occur in anaerobic digesters. These types are increase in alkalinity, increase in carbon dioxide, increase in fatty acids, particulate surfactants, polymers, and temperature fluctuations.

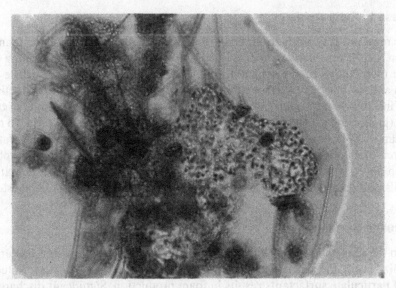

FIGURE 20.9 *Gelatinous material produced by Zoogloeal growth as revealed through Gram staining of a floc particle.*

Increase in Alkalinity

With increasing alkalinity, sludge becomes more surface active and has an increasing propensity to foam. Operational conditions that result in an increase in alkalinity include the following:

- Increased alkalinity loading (ionized ammonia, amino acids, proteins, and cationic polyacrylamide polymers) in the digester feed sludge
- Death and lysis of large numbers of bacteria, especially Nocardioforms, resulting in the release of numerous biosurfactants
- Decrease in the destruction of alkalinity within the anaerobic digester (e.g., decrease fatty acid production)

Increase in Carbon Dioxide

With increasing carbon dioxide production in an anaerobic digester, the quantity of carbon dioxide in the sludge also increases. Increase in carbon dioxide production can occur through increased organic loading to the digester or decrease methanogenesis. The entrapment of carbon dioxide in sludge results in foam production.

Carbon dioxide content within digester sludge can be reduced by bubbling digester gas through a potassium hydroxide (KOH) solution or introducing natural gas into the gas system to dilute carbon dioxide content. A decrease in carbon dioxide content also contributes to an increase in pH and a more favorable volatile acid-to-alkalinity ratio.

Increase in Fatty Acids

An increase in fatty acids in an anaerobic digester also contributes to foam production. This is especially true during start-up operation when acid-forming bacteria are more active than methane-forming bacteria. Fatty acids are surface active agents and decrease the surface tension of sludge resulting in foam production.

An increase in fatty acids may be associated with an increased loading of triglycerides (grease and animal fat) and the death of large numbers of bacteria. To reduce the quantity of fatty acids produced in an anaerobic digester, triglycerides (fats, oils, and grease) should be removed upstream of the digester and treated aerobically with an appropriate bioaugmentation product to ensure adequate degradation of fats, oils, and grease before transferring these wastes to the anaerobic digester.

Particulate Surfactants

Due to the long solids retention time (SRT) of an anaerobic digester, bacteria can solubilize easily the particulate surfactants transferred to the digester. Solubilization of particulate surfactants results in foam production. Significant dischargers of particulate surfactants should be identified and regulated to prevent surfactant foam in an anaerobic digester.

Polymers

Cationic polyacrylamide polymers are used at many wastewater treatment plants for sludge dewatering and sludge thickening. These polymers contain numerous amino groups that are released as the polymer degrades in the anaerobic digester. The released amino groups form ionized ammonia that increase the alkalinity of the digester sludge resulting in foam production. Therefore, periodic testing of polymers to ensure their compatibility for dewatering and thickening sludges should be performed. Also, proper mixing and dosing of polymers should always be practiced to ensure that excess polymer does not enter waste streams.

Temperature Fluctuations

Temperature fluctuations in an anaerobic digester change the dominant and recessive acid-forming bacteria. If the temperature fluctuation favors the proliferation of acid-forming bacteria that produce fatty acids that cannot be used directly or indirectly by methane-forming bacteria for the production of methane, these acids accumulate. The accumulating acids change the surface tension of the sludge and permit foam production. Therefore, the maintenance of a "steady-state" temperature value or narrow range of temperature values should be practiced with anaerobic digesters. The selected temperature or range of temperatures for the operation of the anaerobic digester should favor those acid-forming bacteria that produce fatty acids that are used quickly by methane-forming bacteria.

21

Biological Malodors

The biological (bacterial) production and release of offensive or malodorous compounds in sanitary sewers, lift stations, and wastewater treatment plants are a nuisance to wastewater personnel and people who live near wastewater collection, conveyance, and treatment facilities. Although malodors are not typically addressed by federal and state air quality regulations, they often are regulated by local authorities.

Biological malodor production occurs when wastewater or sludge becomes septic. Septic wastewater or sludge develops when soluble cBOD is degraded through bacterial activity in the absence of free molecular oxygen (O_2) and nitrate (NO_3^-). This degradation of soluble cBOD often is referred to as "septicity." Septicity occurs when wastewater or sludge remains in the sanitary sewer, lift station, or biological treatment unit for an excess time. Examples of where septicity and malodor production and release commonly occur are (1) lift stations and sumps, (2) manholes, (3) sewers, (4) treatment units (headworks, secondary clarifiers, and gravity thickeners, (5) composting operations, (6) dewatering units, and (7) solids handling units (Table 21.1). Biological malodor production and release can occur in an unstable anaerobic digester. The malodors produced in the digester are released from its decant in the headworks of the treatment plant.

Although numerous malodorous compounds are produced through septicity, most malodorous compounds are sulfur-based. These compounds include hydrogen sulfide (H_2S), dimethyl sulfide ((CH_3)$_2$S), and mercaptans. Numerous malodorous compounds also are produced through the anaerobic degradation of carbohydrates, lipids, and proteins or nitrogenous compounds.

Anaerobic degradation of soluble cBOD results in the production of organic compounds and inorganic compounds. Many of these compounds are malodorous, including volatile organic compounds (VOC) and volatile fatty acids (VFA). Some

TABLE 21.1 Wastewater Sites where Septicity and Malodor Production and Release Commonly Occur

Screens and comminutors
  Malodors released from organic wastes adhering to screens
  Malodors released from comminutor wastes adhering to screens
Grit chamber
  Malodors released during cleaning operations
Grit hopper
  Malodors released from stored grit
Wet well
  Malodors released during drawdowns
Preaeration tank
  Malodors produced in sanitary sewer system released during aeration
Clarifiers
  Malodors released from scum on surface
  Malodors released from septic sludge
  Malodors released from biological growth on weirs and walls
Aeration tank
  Malodors carried by aerosols
  Malodors due to inadequate mixing and low dissolved oxygen level
  Malodors produced in accumulated foam
  Volatilization of organic compounds
Air lift (return activated sludge)
  Volatilization of organic compounds from secondary clarifier
Thickener
  Malodors released from septic sludge
Anaerobic digester
  Malodors released from annular space
  Malodors released in decant from unstable digester
Liquid handling
  Malodors released from sludge residues in pipes
  Malodors released during transfer of sludge

inorganic gases also are produced through the anaerobic degradation of soluble cBOD, and some of these gases such as ammonia (NH_3) and hydrogen sulfide are malodorous.

Malodorous compounds are responsible for a number of problems. These problems include malodor complaints, lawsuits, increased operational costs, health risks, concrete and metal corrosion, and the inflow of anaerobic wastewater into activated sludge processes that may trigger undesired filamentous bacterial growth.

Anaerobic degradation of soluble cBOD occurs without free molecular oxygen or nitrate. Although some bacteria use only free molecular oxygen to degrade soluble cBOD (respiration) and many bacteria use free molecular oxygen or nitrate to degrade soluble cBOD (respiration), most bacteria are capable of using another molecule to degrade soluble cBOD. These molecules include sulfate (SO_4^{2-}), an organic molecule, and carbon dioxide (CO_2).

Respiration occurs as bacteria use free molecular oxygen (Equation 21.1) or nitrate (Equation 21.2) to degrade soluble cBOD. When free molecular oxygen is used, carbon dioxide, water, and new bacterial cells are produced. When nitrate ions are used, carbon dioxide, water, molecular nitrogen, and new bacterial cells are produced. No malodorous compounds are produced during the degradation of soluble cBOD when free molecular oxygen or nitrate is used.

$$CH_2O + O_2 \rightarrow CO_2 + H_2O + \text{bacterial cells} \tag{21.1}$$

$$CH_2O + NO_3^- \rightarrow CO_2 + H_2O + N_2 + \text{bacterial cells} \tag{21.2}$$

Forms of anaerobic degradation of soluble cBOD (septicity) that result in the production of malodorous compounds are sulfate reduction and mixed acid and mixed alcohol production (fermentation). Sulfate reduction and fermentation occur in the absence of free molecular oxygen and nitrate or in the presence of an oxygen gradient and nitrate gradient.

When sulfate is used to degrade soluble cBOD, carbon dioxide, water, hydrogen sulfide, sulfide, a mixture of short-chain organic compounds (acids, alcohols, and miscellaneous compounds), and new bacterial cells are produced (Equation 21.3). When an organic molecule is used to degrade soluble cBOD, carbon dioxide, water, a mixture of short-chain compounds (acids, alcohols, and miscellaneous compounds), and new bacterial cells are produced (Equation 21.4). Some of the organic compounds that are produced are volatile organic compounds (VOC), volatile fatty acids (VFA), and volatile sulfur compounds (VSC).

$$CH_2O + SO_4^{2-} \rightarrow CO_2 + H_2O + H_2S + HS^-$$
$$+ \text{organic compounds} + \text{bacterial cells} \tag{21.3}$$

$$CH_2O + CH_2O \rightarrow CO_2 + H_2O + \text{organic compounds} + \text{bacterial cells} \tag{21.4}$$

Volatile fatty acids are short-chain acids that vaporize at atmospheric pressure. Some of these acids are malodorous. If nitrogen-containing and sulfur-containing compounds such as amino acids and proteins are degraded during septicity, nitrogen-containing malodorous compounds and sulfur-containing malodorous compounds or volatile sulfur compounds (VSC) are produced. Also the degradation of nitrogen-containing and sulfur-containing compounds such as the amino acid cysteine ($SHCH_2CHNH_2COOH$) results in the release of malodorous inorganic gases such as ammonia and hydrogen sulfide (Equation 21.5).

$$SHCH_2CHNH_2COOH \xrightarrow{\text{septicity}} CH_3COCOOH + NH_3 + H_2S \tag{21.5}$$

When sulfate reduction occurs, hydrogen sulfide, sulfides, and short-chain acids and alcohols are produced. Although hydrogen sulfide is malodorous, it also is toxicity to nitrifying bacteria. Sulfides serve as an energy source for many sulfide-oxidizing, filamentous organisms such as *Beggiatoa* spp., *Thiothrix* spp., and type 021N. The acids and alcohols produced from sulfate reduction serve as an substrate for many filamentous organisms such as *Nosticola limicola* and type 0041. The undesired growth of these filamentous organisms in activated sludge processes is responsible for settleability problems in the secondary clarifier.

When carbohydrates and lipids are degraded under septicity, numerous short-chain acids and alcohols are produced as well as some miscellaneous compounds

TABLE 21.2 Significant Organic Compounds Produced Through Septicity

Compound	Formula	VOC	VFA	Nitrifying Inhibitor
Acetate	CH_3COOH	X	X	
Acetone	CH_3COCH_3	X		X
Acetaldehyde	CH_3CHO	X		
Butanol	$CH_3(CH_2)_2CH_2OH$	X		
Butyraldehyde	C_2H_5COO	X		
Butyrate	$CH_3(CH_2)_2CH_2COOH$	X	X	
Caproic acid	$CH_3(CH_2)_4COOH$	X	X	
Formaldehyde	CH_2O	X		
Formate	$HCOOH$	X	X	
Ethanol	CH_3CH_2OH		X	X
Lactate	$CH_3CHOHCOOH$			
Methane	CH_4			
Methanol	CH_3OH			X
Propanol	$CH_3CH_2CH_2OH$			X
Propionate	CH_3CH_2COOH	X	X	
Succinate	$CH_3CHOHCOOH$	X	X	
Valeric acid	$CH_3(CH_2)_3COOH$	X	X	

TABLE 21.3 Significant Organic Nitrogen Compounds Produced Through Septicity

Compound	Formula	VOC	Nitrifying Inhibitor
Cadaverine	$H_2N(CH_2)_5NH_2$	X	
Dimethylamine	CH_3NHCH_3	X	
Ethylamine	$C_3H_5NH_2$	X	
Indole	$C_8H_{13}N$	X	
Methylamine	CH_3NH_2	X	
Putresine	$H_2N(CH_2)_4NH_2$	X	
Propylamine	$CH_3CH_2CH_2NH_2$	X	
Pyridine	C_5H_6N	X	
Skatole	C_9H_9N	X	X
Trimethylamine	$CH_3NCH_3CH_3$	X	

(Table 21.2). Many of these organic compounds are VOC and VFA. Several are malodorous, and some inhibit nitrification. When amino acids, proteins, or organic nitrogen compounds are degraded under septicity, a variety of malodorous nitrogen-containing compounds are produced (Table 21.3). If amino acids, proteins, or organic sulfur compounds are degraded under septicity, a variety of malodorous sulfur-containing compounds are produced (Table 21.4). In addition to organic compounds, several inorganic compounds also are produced during anaerobic degradation of soluble cBOD (Table 21.5).

When degradation of soluble cBOD occurs, electrons are transferred from the degrading organic compound to an electron transport molecule. The transport molecule may be free molecular oxygen, nitrate, sulfate, an organic molecule, or carbon dioxide. The transfer of electrons is known as an oxidation–reduction reaction. Within the wastewater or sludge the molecule that is being used to transport electrons can be determined by the oxidation–reduction potential (ORP) of the

TABLE 21.4 Significant Organic Sulfur Compounds Produced Through Septicity

Compound	Formula	VSC
Allyl mercaptan	$CH_2=CHCH_2SH$	X
Benzyl mercaptan	$C_6H_5CH_2SH$	X
Dimethyl sulfide	$(CH_3)_2S$	X
Dimethyl disulfide	CH_3SSCH_3	X
Ethyl mercaptan	C_2H_5SH	X
Methyl mercaptan	CH_3SH	X
Thiocresol	$CH_3C_6H_4SH$	X
Thioglycolic acid	$HSCH_2COOH$	X

TABLE 21.5 Significant Inorganic Compounds Produced Through Septicity

Name	Formula	Nitrifying Inhibitor
Ammonia	NH_3	
Carbon dioxide	CO_2	
Carbon disulfide	CS_2	
Carbon monoxide	CO	
Hydrogen sulfide	H_2S	X
Molecular nitrogen	N_2	
Nitrous oxide	N_2O	
Water	H_2O	

TABLE 21.6 Guideline of Ranges of ORP Values and Use of Final Electron Transport Molecules

ORP (mV)	Final Electron Transport Molecule	Form of Degradation
>+200	O_2	Aerobic, nBOD and cBOD removal
>+100	O_2	Aerobic, cBOD removal
+100 to −100	NO_3^-	Anoxic, cBOD removal
<−100	SO_4^{2-}	Anaerobic (sulfate reduction), cBOD removal
<−200	Organic molecule	Anaerobic (mixed acid/mixed alcohol production) cBOD removal
<−300	CO_2	Anaerobic, methane production

wastewater or sludge. The ORP is measured in millivolts (mV) and may be positive or negative (Table 21.6).

CONTROL OF BIOLOGICAL MALODORS

Although the use of ORP is helpful in locating sites of septicity or malodor production, ORP does not correct malodor production. Control of biological malodor production can be achieved through several biological, chemical, and physical measures (Table 21.7).

TABLE 21.7 Operational Measures Available for Control of Biological Malodor Production

Anoxic transformation
Bioaugmentation
Biofilters and biotowers
Chemical treatment
Cleaning
Design
pH adjustment
Reaeration
Regulation of industrial discharges

Anoxic Transformation

In the absence of free molecular oxygen and nitrate, nitrate may be added to the sanitary sewer system or biological treatment unit that is experiencing septicity. The addition of nitrate produces an anoxic condition. Biological malodor production cannot occur in the presence of nitrate or when the ORP of the wastewater or sludge is $>-100\,mV$; that is, sulfate reduction and mixed acid and mixed alcohol production cannot occur. Nitrate can be added in the form of calcium nitrate ($Ca(NO_3)_2$) or sodium nitrate ($NaNO_3$).

Caution must be exercised when adding nitrate. Excess nitrate in the sanitary sewer system permit the rapid degradation of soluble cBOD. This depletion reduces the cBOD concentration of the wastewater. With a smaller concentration of cBOD in the influent to a wastewater treatment plant, it becomes more difficult for the wastewater treatment plant to achieve its required removal percentage for cBOD.

Bioaugmentation

Bioaugmentation is the addition of highly concentrated, commercially prepared bacterial cultures to the sanitary sewer system or biological treatment unit. Some bacteria such as *Pseudomonas* spp. have the enzymatic ability to degrade specific malodorous compounds before these compounds are released to the atmosphere.

Biofilter and Biotowers

Biofilters and biotowers are solid media systems that permit the growth of a large and diverse population of bacteria on an organic-based media such as bark, compost, or wood chips. The bacteria grow as a biofilm on the media and degrade the malodorous compounds as they are absorbed to the film. Gases released from treatment units are collected and vented through biofilters or biotowers.

In biofilters the nutrients needed by the bacteria are provided in the media. In biotowers the nutrients needed by the bacteria are provided in a water spray or added to the recirculation water.

Chemical Treatment

Several chemicals may be added to the sewer system to control the biological production of malodorous compounds. Chemicals commonly used include sodium hydroxide (NaOH), calcium hydroxide ($Ca(OH)_2$), chlorine as sodium hypochlorite (NaOCl) or calcium hypochlorite ($Ca(OCl)_2$), hydrogen peroxide (H_2O_2), ozone (O_3), and metal salts.

By increasing the pH of the sewer system to >12 with sodium hydroxide or calcium hydroxide, the biofilm within the sewer system degrades. With the degradation of the biofilm or disinfection of bacteria, the biological production of malodorous compounds is temporarily stopped. With the regrowth of the biofilm, biological production of malodorous compounds may again occur.

The use of chlorine to control the biological production of malodorous compounds has biological and chemical impacts. Biologically, chorine disinfects. Chemically, chlorine oxidizes malodorous compounds such as sulfides (HS^-) to sulfur (S^0). The oxidation of many malodorous compounds renders them less offensive or nonoffensive.

Hydrogen peroxide and ozone also disinfect. These chemicals quickly destroy sulfate-reducing bacteria in the biofilm that are responsible for the production of hydrogen sulfide (H_2S). Hydrogen peroxide also oxidizes sulfides to sulfate ions (SO_4^{2-}).

Metal salts, such as chlorides of iron, may be added to the sewer system to precipitate sulfides from solution. The use of metal salts results in the formation of insoluble iron sulfide (FeS).

Cleaning

Thorough and periodic cleaning of problematic areas of the sewer system or treatment units helps to remove biofilm, sediment, or solids that permit septic conditions. Hydraulic flushing and winching of sewers may be needed in order to prevent septicity.

Design

Sewer systems should be designed to reduce the amount of turbulence of anaerobic wastewater and the release of malodorous compounds and the amount of solids that accumulate. Treatment units also should be designed to ensure adequate removal of settled solids and prevent the accumulation of solids in corners or sides of treatment tanks. Short-circuiting of flow patterns should be prevented and if needed, baffles should be install to correct for short-circuiting.

Ventilation of sewers and treatment units and the treatment of vented gases should be considered. Ventilation also reduces the amount of moisture on sewer walls and treatment unit walls. The reduction in moisture helps to reduce septicity.

pH Adjustment

The release of hydrogen sulfide and ammonia to the atmosphere is strongly pH-dependent. It is the non-ionized forms of hydrogen sulfide (H_2S) and ammonia (NH_3) that are released to the atmosphere. The ionized forms of hydrogen sulfide (HS^-) and ammonia (NH_4^+) are not released to the atmosphere. These forms remain in solution. However, with decreasing pH, H_2S is produced and released to the atmosphere. With increasing pH, NH_3 is produced and released to the atmosphere.

Therefore, it is important to maintain a near-neutral pH in the sewer system or treatment unit when either of these inorganic compounds is present in order to prevent the release of the compound to the atmosphere.

Reaeration

By increasing the dissolved oxygen level in the sewer system or treatment unit, septicity is prevented. Reaeration can be achieved through the injection of air or pure oxygen. Sulfate reduction or the production of hydrogen sulfide usually does not occur in gravity sewers when the dissolved oxygen concentration is >0.5 mg/liter.

Regulation of Industrial Discharges

Many industrial wastes contain malodorous compounds such as VOC. Dischargers of malodorous compounds should be identified and regulated.

22

Atmospheric Inversions

A major problem associated with the operation of wastewater treatment plants is the microbial (bacterial) production and release of malodors. Even the most efficiently operated plants experience malodor problems, and they are a concern to wastewater personnel and people who live near the treatment plants. These problems involve economic, legal, technical, and public relations issues. The intensity of the malodorous condition may be greatly enhanced, if a wastewater treatment plant experiences an adverse weather condition such as an atmospheric inversion or unfavorable topography.

ATMOSPHERIC INVERSION

An atmospheric inversion is an adverse weather condition that can intensify malodors and malodor-related problems. Atmospheric inversions most commonly occur during warm and often humid summer months.

During a cloudless summer day, the ground is warmed directly by the sun and more quickly than the atmosphere. The atmosphere is then warmed indirectly by the upward movement of heat or "thermals" from the ground. With the upward movement of thermals, malodors that are produced at a wastewater treatment plant are carried into the atmosphere (Figure 22.1). This atmospheric condition reduces the movement of malodors from the wastewater treatment plant to adjacent neighborhoods.

During a cloudless night, the ground loses heat more quickly than the air. The air closest to the ground cools most rapidly. This change or inversion of a warm "blanket" of air during the day to a cool "blanket" of air during the early evening hours (6–8 P.M.) entraps malodors in the cool blanket of air (Figure 22.2). This

Wastewater Bacteria, by Michael H. Gerardi

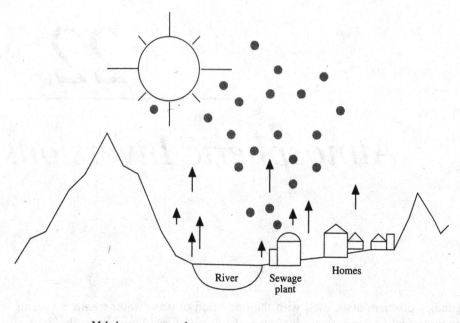

● Malodorous compound

↑ Thermal

FIGURE 22.1 *Atmospheric inversions, thermals. During the day the ground warms more quickly than the atmosphere. This results in the production of thermals that carry malodorous compounds away from the wastewater treatment plant and nearby neighborhoods.*

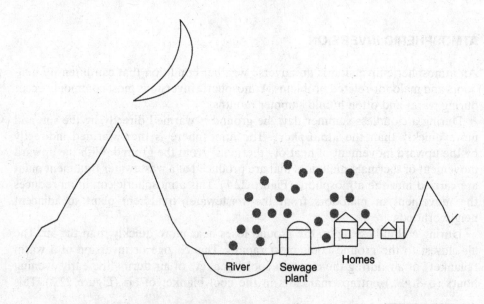

● Malodorous compound

FIGURE 22.2 *Atmospheric inversions, inversion. During the evening hours the air cools more quickly than the ground. This results in the entrapment of malodorous compounds immediately above the wastewater treatment plant and nearby neighborhoods.*

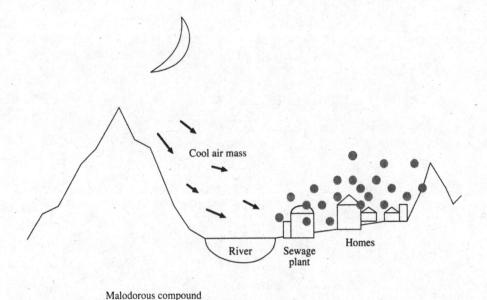

FIGURE 22.3 *Atmospheric inversions, dispersion. If hills or mountains are near the wastewater treatment plant, a cool air mass may move down on the wastewater treatment plant and push the blanket of malodorous compounds into more distant neighborhoods.*

blanket usually is from the ground level to approximately 100 feet above the wastewater treatment plant.

As the malodors increase in concentration within the cool blanket of air, they began to diffuse away from the wastewater treatment plant into adjacent neighborhoods. The offensive nature of the malodors is enhanced with the presence of high humidity.

UNFAVORABLE TOPOGRAPHY

Malodors may be dispersed into a larger number of neighborhoods due to unfavorable topography. If the wastewater treatment plant is located near elevated topography such as hills or mountains, cool air from the elevated topography may descend upon the plant and disperse the entrapped malodors into more distant neighborhoods (Figure 22.3).

The cool air descending upon the wastewater treatment plant is produced from (1) the elevation of the topography and (2) the presence of a vegetative canopy. These two conditions permit a more rapid cooling of the air mass immediately above the wastewater treatment plant. Wastewater treatment plants that are located adjacent to undesired topography and experience atmospheric inversions are at more risk to malodor-related problems than those plants that do not have topographic concerns and the occurrence of atmospheric inversions.

References

Al-Shahwani, S. M., and N. J. Horan. 1991. The use of protozoa to indicate change in the performance of activated sludge plants. *Water Res.* (25).

Anderson, J. W. 1978. *Sulphur in Biology*. Univerity Park Press, Baltimore.

Bonnin, C., A. Loborie, and H. Paillard. 1990. Odor nuisances created by sludge treatment: problems and solutions. *Water Sci. Tech.* (22).

Britton, G. 1994. *Wastewater Microbiology*. Wiley-Liss, New York.

Doetsch, R. N., and T. M. Cook. 1973. *Introduction to Bacteria and Their Ecobiology*. University Park Press, Baltimore.

Dold, P. L., G. A. Ekama, and G. R. Marais. 1980. A general model for the activated sludge process. *Prog. Water Tech.* (12).

Easter, C., C. Quigley, P. Burrowes, J. Witherspoon, and D. Apgar. 2005. Biotechnology gains ground as a tool for odor and air emissions control. *Water Env. Tech.* (5).

Eikelboom, D. H. 1994. The *Microthrix parvicella* puzzle. *Water Sci. Tech.* (29).

Fry, J. C., G. M. Gadd, R. A. Herbert, C. W. Jones, and I. A. Watson-Craik, eds. 1992. *Microbial Control of Pollution*. Society of General Microbiology, Cambridge University Press, London.

Gerardi, M. 2005. *Wastewater Pathogens*. Wiley-Interscience, New York.

Gerardi, M. 2003. *The Microbiology of Anaerobic Digesters*. Wiley-Interscience, New York.

Gerardi, M. 2002. *Nitrification and Denitrification in the Activated Sludge Process*. Wiley-Interscience, New York.

Gerardi, M. 2002. Taming sewer smells; biological malodor production and control in sewer systems. *Env. Protection* (8).

Gerardi, M. 2002. *Settleability Problems and Loss of Solids in the Activated Sludge Process*. Wiley-Interscience, New York.

Gerardi, M. 1981. Treatment plant odors and atmospheric inversion conditions. Deeds & Data, Supplement to *WPCF Highlights*.

Gottschalk, G. 1979. *Bacterial Metabolism*. Springer-Verlag, New York.

Hvited-Jacobsen, T. 2002. *Sewer Processes: Microbial and Chemical Process Engineering of Sewer Newtworks*. CRC Press, Boca Raton, FL.

Jenkins, D., M. G. Richard, and G. T. Daigger. 1993. *Manual on the Causes and Control of Activated Sludge Bulking and Foaming*, 2nd ed. Lewis Publishers, Boca Raton, FL.

Kadlec, R. H., and R. L. Knight. 1996. *Treatment Wetlands*. CRC Press, Boca Raton, FL.

Lankford, P. W., and L. Smith. 1994. Toxicity measurements and their use in treatment strategies, *J. IWEM*. (8)

Lawrence, A. W., and P. L. McCarty. 1965. The role of sulfide in preventing heavy metal toxicity in anaerobic treatment. *J. Water Poll. Control Fed*. (37).

Leschine, S. B. 1995. Cellulose degradation in anaerobic environements. *Annu. Rev. Microbiol*. (49).

Liao, B. Q., D. G. Allen, I. G. Droppo, et al. 2001. Surface properties of sludge and their role inbioflocculation and settleability. *Water Res*. (35).

Mara, D., and N. Horan, eds. 2003. *Handbook of Water and Wastewater Microbiology*. Academic Press, New York.

McCarty, P. L., and D. P. Smith. 1986. Anaerobic wastewater treatment. *Env. Sci. Tech*. (20).

McCarty, P. L., and D. P. Smith. 1964. Anaerobic waste treatment fundamentals, Part III: toxic materials and their control. *Public Works* (95).

McCarty, P. L. 1964. Anaerobic waste treatment fundamentals, Part II: environmental requirements and control. *Public Works* (95).

McCarty, P. L., and R. E. McKinney. 1961. Volatile acid toxicity in anaerobic digestion. *J. Water Poll. Control Fed*. (33).

Mizuno, O., Y. Y. Li, and T. Noike. 1998. The behavior of sulphate-reducing bacteria in acidogenic phase of anaerobic digestion. *Water Res*. (32).

Mudrack K., and S. Kunst. 1986. *Biology of Sewage Treatment and Water Pollution Control*. Ellis Horwoood Limited, Chichester, England.

Postgate, J. R. 1984. *The Sulfate Reducing Bacteria*, 2nd ed. Cambridge University Press, Cambridge, England.

Ramanathan, M., and A. F. Gaudy, Jr. 1972. Sludge yields in aerobic systems. *J. Water Poll. Control Fed*. (44).

Schmitt, F. and C. F. Seyfried. 1992. Sulphate reduction in sewer sediments. *Water Sci. Tech*. (25).

Speece, R. E. 1983. Anaerobic wastewater treatment. *Env. Sci. Tech*. (9).

Standard Methods. 1985. *Standard Methods for the Examination of Water and Wastewater*, 16th ed. Prepared and published jointly by American Public Health Association, American Water Works Association, and Water Pollution Control Federation. Washington, DC.

Taylor, G. T. 1982. The methanogenic bacteria. *Prog. Ind. Micro*. (16).

Trevors, J. T., and C. M. Cotter. 1990. Copper toxicity and uptake in microorganisms. *J. Ind. Micro*. (6).

Traxler, R. W., and E. M. Wood. Bioaccumulation of metals by coryneform SL-1. *J. Ind. Micro*. (6).

U.S. EPA. 1985. *Odor and Corrosion Control in Sanitary Systems and Treatment Plants*. Report #EPA/625/1-85-018.

Warren, R. A. J. 1996. Microbial hydrolysis of polysaccharides. *Annu. Rev. Microbiol*. (50).

Wentzel, M. C., G. A. Ekama, and G. R. Marais. 1992. Processes and modelling of nitrification denitrification biological excess phosphorus removal systems—A review. *Water Sci. Technol* (25).

White, D. 2000. *The Physiology and Biochemistry of Prokaryotes*. Oxford University Press, New York.

Zehnher, A. B. J. 1988. *Biology of Anaerobic Microorganisms*. John Wiley and Sons, New York.

Zeikus, J. G. 1977. The biology of methanogenic bacteria. *Bacteriol. Rev.* (41).

Abbreviations and Acronyms

ABS	Alkylbenzene sulfonate
ATP	Adenosine triphosphate
BOD	Biochemical oxygen demand
°C	Degrees Celsius
cBOD	Carbonaceous biochemical oxygen demand
cm	Centimeter
COD	Chemical oxygen demand
DO	Dissolved oxygen
DNA	Deoxyribonucleic acid
EBPR	Enhanced biologicl phosphorus removal
e.g.	*exempli gratia*, for example
F/M	Food-to-microorganism ratio
GAC	Granular activated carbon
HRT	Hydraulic retention time
i.e.	*id est*, that is
I/I	Inflow and infiltration
LAS	1-benzenesulfonate
lb	pound
MCRT	Mean cell residence time
mg	Milligram
mg/liter	Milligram per liter
MLSS	Mixed liquor suspended solids
MLVSS	Mixed liquor volatile suspended solids

Wastewater Bacteria, by Michael H. Gerardi
Copyright © 2006 John Wiley & Sons, Inc.

MSDS	Material safety data sheet
mV	Millivolt
nBOD	Nitrogenous biochemical oxygen demand
ORP	Oxidation-reduction potential
PAO	Phosphorus accumulating organism
PHB	Poly-β-hydroxybutyrate
μΩ	Micro-ohm
μm	Micron
RAS	Return activated sludge
RNA	Ribonucleic acid
SOUR	Specific oxygen uptake rate
SRB	Sulfate-reducing bacteria
SRT	Solids retention time
TKN	Total kjeldahl nitrogen
TSS	Total suspended solids
VFA	Volatile fatty acid
VOC	Volatile organic compound
VSC	Volatile sulfur compound
VSS	Volatile suspended solids

Chemical Compounds and Elements

Ag	Silver
Al	Aluminum
Al^{2+}	Aluminum ion
As	Arsenic
C	Carbon
Ca	Calcium
Ca^{2+}	Calcium ion
$CaCO_3$	Calcium carbonate
$Ca(HSO_3)_2$	Calcium bisulfite
$Ca(OCl)_2$	Calcium hypochlorite
$Ca(OH)_2$	Calcium hydroxide
$CaOH(PO_4)_3$	Hydroxyapatite
$Ca(NO_3)_2$	Calcium nitrate
CCl_4	Carbon tetrachloride
CdS	Cadmium sulfide
$—CH_3$	Methyl group
CH_4	Methane
$CH_3CH_2CH_2OH$	Isopropanol, n-propanol
$CH_3CH_2CH_2COOH$	Butyrate
$C_6H_{12}O_6$	Glucose
CH_3CH_2OH	Ethanol
$(CH_3)_2CHOH$	i-Propanol
$CH_2(CH_3)_3COOH$	Valeric acid

Wastewater Bacteria, by Michael H. Gerardi
Copyright © 2006 John Wiley & Sons, Inc.

CH_3CH_2COOH	Propionate
$CH_3CHOHCOOH$	Lactate
$CH_2=CHCH_2SH$	Allyl mercaptan
$CH_3CH=CHCH_2SH$	Crotyl mercaptan
$CH_3CH_2CH_2CH_2SH$	Amyl mercaptan
CH_3CH_2SH	Ethyl mercaptan
$(CH_3)_3CSH$	t-Butyl mercaptan
CH_3CCl_3	1,1,1-Trichloroethane
CH_2Cl_2	Methylene chlorine
CH_3Cl_3	Chloroform
$CH_3COC_2H_5$	Ethyl acetate
$CH_3COCOOH$	Pyruvate
$(CH_3)_3COH$	t-Butanol
CH_3COOH	Acetate
CH_3NH_2	Methylamine
C_8H_6NH	Indole
C_9H_8NH	Skatole
$(CH_3)_2NH_2$	Dimethlyamine
$CH_3NH_2CH_2OH$	Amionoethanol
CH_2NH_2COOH	Glycine
CH_3OH	Methanol
$C_5H_7O_2N$	Cellular material
$(CH_3)_2S$	Methylsulfide
$(CH_3)_2S_2$	Methyldisulfide
CH_3SH	Methy thiol, methylmercaptan
C_2H_9SH	Ethylmercaptan
CH_3SSCH_3	Dimethyl disulfide
Cl	Chlorine
Cl^-	Chloride ion
Cl_2	Gaseous chlorine
CN^-	Cyanide
Co	Cobalt
Co^{2+}	Cobalt ion
CO	Carbon monoxide
CO_2	Carbon dioxide
CO_3^{2-}	Carbonate
$-COOH$	Carboxyl group
Cu^{2+}	Copper ion
CuO	Copper oxide
Cr^{3+}	Trivalent chromium
Cr^{6+}	Hexavalent chromium
F	Fluoride
Fe	Iron
Fe^{2+}	Ferrous
Fe^{3+}	Ferric
$Fe(CN)6^{4-}$	Ferrocyanide
FeS	Iron sulfide
H	Hydrogen

H^+	Hydrogen proton
H_2	Hydrogen gas
HCN	Hydrogen cyanide
HCO_3^-	Bicarbonate alkalinity
H_2CO_3	Carbonic acid
HCOOH	Formate
Hg	Mercury
H_2NCONH_2	Urea
HNO_3	Nitrous acid
H_2O	Water
H_2O_2	Hydrogen peroxide
HOCl	Hypochlorous acid
$HOOCCH_2CH_2COOH$	Succinate
HOOCCHCOOH	Fumarate
HPO_4^{2-}	Orthophosphate
$H_2PO_4^-$	Orthophosphate
HS^-	Sulfide
H_2S	Hydrogen sulfide
HSO_3^-	Hydrogen sulfite
H_2SO_3	Sulfurous acid
K	Potassium
K^+	Potassium ion
$KMnO_4$	Potassium permanganate
KOH	Potassium hydroxide
Mg	Magnesium
Mg^{2+}	Magnesium ion
$Mg(HSO_3)_2$	Magnesium bisulfite
Mn	Manganese
Mo	Molybdenum
N	Nitrogen
N_2	Molecular or gaseous nitrogen
Na	Sodium
Na^+	Sodium ion
$-NH_2$	Amino group
NH_3	Ammonia
NH_4^+	Ionized ammonia
Ni	Nickel
NaOCl	Sodium hypochlorite
NaOH	Sodium hydroxide
Na_2SO_3	Sodium sulfite
NO	Nitric oxide
N_2O	Nitrous oxide
NO_2^-	Nitrite
NO_3^-	Nitrate
O	Oxygen
O_2	Free molecular oxygen
O_2^-	Superoxide
O_3	Ozone

OCl^-	Hypochlorous ion
—OH	Hydroxly group, hydroxyl ion
P	Phosphorus
Pb	Lead
PO_4^{3-}	Orthophosphate
$P_2O_7^{3-}$	Polyphosphate, pyrophosphate
S	Sulfur
S_2^-	Sulfide
S°	Elemental sulfur
=S—C—	Thioketone
—SH	Thiol group
$SHCH_2CHNH_2COOH$	Cysteine
=S—O—	Sulfoxide
SO_3^{2-}	Sulfite
SO_4^{2-}	Sulfate
$S_2O_3^-$	Thiosulfate
$S_2O_4^{2-}$	Dithionite
$S_2O_5^{2-}$	Disulfite
S_2O_6	Dithionate
—SOOOH	Sulfonic acid group
—S—S—	Disulfide bonds
Zn^{2+}	Zinc ion
ZnS	Zinc sulfide

Glossary

abiotic The nonliving components or factors in an environment that affect an organism

absorb Penetration of a substance into the body of an organism

acclimate The process by which bacteria spend time and energy to repair damage to enzyme systems caused by toxic wastes

acetoclastic The splitting of acetate by methane-forming bacteria to produce methane

acetongenic Acid-forming or fermentative bacteria that produce large quantities of acetate

acid-forming bacteria Organisms that produce a mixture of acids, alcohols, and other compounds during anaerobic or fermentative conditions

acidophile An organism that lives under acidic conditions

actinomycete A group of bacteria that are Gram positive and highly branched and that also share some characteristics of fungi; responsible for settleability problems and foam production in the activated sludge process

active transport The movement of substrates and nutrients from the bulk solution to the cytoplasm that requires an expenditure of energy by an organism

activator A component of some enzymes that improves enzymatic activity

acute toxicity Sudden and short-term onset of toxicity

adsorb The taking up of one substance at the surface of an organism

adjustment period Time required by bacteria to produce enzymes for the degradation of substrate

aerobe An organism that uses free molecular oxygen for the degradation of substrate

Wastewater Bacteria, by Michael H. Gerardi
Copyright © 2006 John Wiley & Sons, Inc.

aerotaxis   Movement of an organism toward oxygen

aerotolerant   An anaerobe that can survive the presence of free molecular oxygen

agglutinate   Process by which floc-forming bacteria stick together or flocculate

alkalinophile   An organism that lives under alkaline conditions

alky chain   A carbon compound with two or more carbon units joining in chain-like fashion

amino acid   A group of organic acids in which a hydrogen atom of the hydrocarbon (alkyl) radical is exchanged for the amino group; used in the production of proteins

anabolism   Biochemical reactions that result in the production of complex substances from simpler ones

anaerobe   An organism that uses a molecule other than free molecular oxygen to degrade substrate

anoxic   A condition that contains nitrate but no free molecular oxygen

antagonistism   The interference of one waste upon the toxicity effect of another waste

archaebacteria   Group of ancient bacteria that includes the methane-forming bacteria

aromatic compound   A benzene derivative or ring compound containing double bonds

autotroph   An organism that obtains its carbon for cellular synthesis from inorganic carbon or carbon dioxide

axial filament   The rigid structure or flagellum of spirochetes that extends through the organism and is responsible for locomotion

bacillus   Rod-shaped cells

binary fission   Typical means of reproduction in bacteria where cells simply split in half

bioaerosol   An aerosol that contains viruses, bacteria, or fungi that are released at wastewater treatment plants

bioaugmentation   The addition of commercially prepared cultures of bacteria to a wastewater treatment plant to improve operational conditions

biomass   The quantity or weight of all organisms within the treatment process

biosurfactant   Surface-active agents released by living or dead cells that contribute to foam production in a wastewater treatment plant

biotic   The living components in an environment

biotin   A vitamin B complex

budding   Mode of reproduction use by fungi such as yeast and some bacteria

capsule   A protective structure that surrounds the cell. It is almost always composed of polysaccharides. The capsule protects the bacterium from harsh environments.

carrying capacity   The maximum number of species or organisms that an environment can support

hydrolysis   Splitting of molecules using water

hydrolytic bacteria   Bacteria that add water to complex molecules and split them into smaller and soluble molecules

indigenous   Native

inhibition   Loss of enzymatic activity due to a toxic waste

inorganic   Compounds that do not contain carbon and hydrogen

isomer   Compound having more than one molecular structure

invasion   Entrance of a pathogen into a host through an abrasion or cut

ligand   An inorganic or organic molecule in solution that holds a metal in solution

lipase   An enzyme that solubilizes lipids

lithotroph   An organism that obtains its energy for the oxidation of a mineral such as nitrogen or sulfur

lysis   Splitting or breaking open of a dead bacterial cell

metabolism   The sum of all chemical processes carried out by a cell

metazoa   Multicellular animals that may be microscopic or macroscopic in size

methanogens   Methane-forming bacteria

methemoglobinemia   Often referred to as "blue baby" disease; occurs from the consumption of potable water contaminated with nitrate

methyltroph   Methane-forming bacteria that use methyl groups ($—CH_3$) to produce methane

mineralization   Chemical or biological breakdown of insoluble minerals (nitrogen, phosphorus, and sulfur) to soluble forms

morphology   Structure or form

niche   Role an organism performs in an environment

nitrification   Biological oxidation of ionized ammonia to nitrite and/or biological oxidation of nitrite to nitrate

nocardioform   Actinomycete

oocyst   Dormant protective stage of protozoa such as *Crytosporidium*

organic   Compound that contains carbon and hydrogen

organotroph   An organism that obtains its carbon and energy from organic compounds

pathogen   Disease-causing agent

peptide bond   Bond joining one amino acid to another

phototaxis   Movement of an organism toward sunlight

phototroph   An organism that obtains its energy from sunlight

phytin   Phosphorus-containing acid found in green leafy vegetable

poly-P bacteria   Bacteria that remove phosphorus from the environment in quantities greater than cellular requirements

procaryote   An organism that contains no nucleus or other membrane-bound organelles

protease   An enzyme that solubilizes proteins

readily available   Nutrients that are in the proper oxidation state or valence for immediate used by cells for enzymatic and synthetic processes

rumen   A separate compartment in the digestive tract of animals such as cows that contain methane-forming bacteria

saprophyte   An organism that lives off dead or decaying organisms

septicity   Bacterial degradation of soluble substrate in the absence of free molecular oxygen and nitrate

sequester   To hold a soluble metal in solution

soaponification   Hardening of lipids

spirillum   Spiral-like shape

strict   Obligate

substrate   Source of carbon or energy for used by an organism

symbiotic   Beneficial relationship between two or more organisms

synergistic   The ability of one toxic substance to increase the toxic effect or another

taxis   Movement or attraction to

tetrad   Group of four

thermoacidophile   An organism that lives in a hot and acidic environment

thiamin   A vitamin B complex

trophic   Food or substrate level

volutin   Polyphosphate granules in poly-P bacteria

Index

abiotic factors 11, 14, 18
Acetobacter 36, 38, 219, 220
Acetobacteracae 35
acetogenic bacteria 12, 13, 14, 18, 35, 159, 177
Achromobacter 37, 39, 93, 110, 136
Acinobacter 22, 38, 39, 93, 110
Actinomadura 39
Actinomyces isrelii 50
actinomycetes 39
adjustment period 42, 68,
Aerobacter 37, 39, 43, 110, 136, 158
Aeromonas 38, 110, 159
Agrobacterium 93
Alcaligenes 16, 37, 39, 93, 136
algae 103, 143
ammonia 14, 16, 18, 25, 27, 34, 38, 39, 44, 66, 77, 78, 79, 81, 83, 84, 85, 86, 87, 88, 94, 95, 97, 100, 113, 115, 162, 169, 175, 179, 180, 181, 182, 183, 186, 187, 193, 194, 195, 196, 200, 215, 218, 221, 222, 223, 224, 229
Amoebobacter 126
Anabaena 36
anabolism 55, 57, 70 -73
Anacalochloris 126
anoxic 16, 66, 72, 73, 89, 91, 97, 99, 100, 101, 115, 116, 149, 150, 151, 156, 205, 207, 219, 228

A/O process 113, 114
A²/O process 113, 114, 115
archaebacteria 3, 4, 10
Archaea 10, 161
ATP 103, 104
Arthrobacter 39, 106, 110, 126, 136
Aspergillus 106, 207
atmospheric inversions 231–233
autotrophs 25, 34, 85

Bacillus 22, 30, 37, 39, 43, 50, 93, 106, 110, 126, 136, 158, 200
Bacteroides 37, 38, 159
Bardenpho process 113, 114, 115
batch culture 67, 68, 70
Beggiatoa 22, 38, 39, 40, 110, 122, 123, 125, 126, 127, 129, 144, 145, 160, 172, 225
bioaugmentation 41–47, 83, 171, 207, 228
biogas 184, 186
Bifidobacteria 37, 38, 159
binary fission 20, 65, 67
biological holdfast system 83
biological nutrient removal (BNR) 113, 116
bioaerosols 49
biochemical oxygen demand (BOD) 15, 47, 72, 73, 169, 172, 186, 188, 191, 192
biofilm 24, 36, 107, 122, 143, 229
bioindicators 189